THE WHOLE WHALE CATALOG

Edited by Barbara Brooks

Greatlakes Living Press, Ltd., Publishers, Waukegan, Illinois

The Whole Whale Catalog

Printed in U.S.A.
Library of Congress Catalog Card No.: 77-846555
International Standard Book Number: 0-915498-62-6

Greatlakes Living Press, Ltd.
91 N. Noll Road
Waukegan, Illinois 60085

ACKNOWLEDGMENTS

Special thanks to Kate Bouve', Director of Information for the Center for Environmental Education, who steered me in many helpful directions. NOAA, the U.S. Marine Mammal Commission and the New York Zoological Society also gave helpful suggestions for finding things.

Thanks also to my daughter Elizabeth, who criticized selections and looked up things I didn't know.

CONTENTS

CONTENTS

SOURCES: Whale Watchers

Charles M. Scammon—The Marine Mammals of the North-Western Coast of North America (San Francisco, 1874; reprinted 1968 by Dover Publications, Inc.)

King Alfred the Great—"The Voyage of Octher," as recounted by King Alfred in his Description of Europe , and translated by Hakluyt in his Principal Navigations.

Hamlet, Prince of Denmark—William Shakespeare, Hamlet.

Thor Heyerdahl—Kon-Tiki (Rand McNally, 1950).

Pliny the Elder (Caius Plinius Secundus)—Historia Naturalis (Natural History), translated by H. Rackham.

Henry David Thoreau: Cape Cod (published posthumously, 1865).

Job: The Bible.

Victor Scheffer: Quoted from "Exploring the Lives of Whales, National Geographic , December 1976.

IN THE BEGINNING

And God said, Let the waters bring forth abundantly the moving creature that hath life, and fowl that may fly above the earth in the open firmament of heaven. And God created great whales, and every living creature that moveth, which the waters brought forth abundantly, after their kind. . . and God saw that it was good.

—*The Bible*

. . . the moot point is, whether Leviathan can long endure so wide a chase and so remorseless a havoc; whether he must not at last be exterminated from the waters, and the last whale, like the last man, smoke his last pipe, and then himself evaporate in the final puff.

—Herman Melville, *Moby Dick*

And On the Fourth Day. . .

Forthwith the sounds and seas, each creek and bay,
With fry innumerable swarm, and shoals
Of fish, that with their fins and shining scales
Glide under the greenwave, in sculls that oft
Bank the mid sea: part single, or with mate
Graze the seaweed their pasture, and through groves
Of coral stray, or sporting with quick glance
Show to the sun their waved coats dropt with gold;
Or in their pearly shells at ease attend
Moist nutriment, or under rocks their food
In jointed armour watch: on smooth the seal
and bended dolphins play; part huge of bulk,
Wallowing unwieldy, enormous in their gait
Tempest the ocean: there Leviathan,
Hugest of living creatures, on the deep
Stretch'd like a promontory sleeps, or swims
And seems a moving land, and at his gills
Draws in, and at his trunk spouts out a sea.

—John Milton
Paradise Lost

1
Beluga whale and her calf in the St. Lawrence — estuary. (Photo courtesy of Jean Laurin)

1

THE WORLD OF THE WHALE

And God created great whales, and every living creature that moveth, which the waters brought forth abundantly. . . .

Genesis
—*The Bible*

The Little Calf: September

Victor Scheffer

The Little Calf, whose life is chronicled so sensitively in The Year of the Whale, *is a sperm whale. As this passage describes, the baby sperm whale is "precocious" in the scientific sense—well-developed, alert, able to swim. He probably weighs about a ton and is 14 feet long; his mother weighs perhaps sixteen times as much and is about 38 feet long.*

2
Narwhal and her calf. (Photo courtesy of Randall R. Reeves)

It is early September when for the first time the Little Calf sees light—a blue-green, dancing light. He slips easily from his mother's body beneath the surface of the Pacific Ocean two hundred miles west of Mexico, on the Tropic of Cancer. He trembles, for the water is cold, and has lain for sixteen months in a warm chamber at ninety-six degrees. He gasps for air as his mother nudges him anxiously to the surface with her broad snout. He breathes rapidly and desperately for a while, puffing with each breath a small cloud of vapor down the autumn breeze.

The sky above vibrates with a piercing clarity that we who live in the shadow of city smoke have long forgotten. Scattered over the riffled surface of the water as far as the eye can see are small groups of whales. A black back rolls slowly in the sun and glistens for

---

The Little Calf

an instant as it disappears. Vapor plumes rise and dissolve in the distance. An occasional *thwack* like a rifle shot breaks the calm, its source a mystery, perhaps a love pat or a mock battle between cavorting youngsters. No bird is in the air, but suddenly, as if by magic, an albatross appears and drops heavily to the water, tearing the surface for twenty yards with its broad webbed feet. It has sighted a fresh, purple-red placenta. Here, on the calving grounds of the sperm whale, the season of birth is at its height.

Whales share with sea cows (manatees and dugongs) and with the hippopotamus of Africa the distinction of being the only mammals born under water. The Little Calf is born tail first, like all young whales, though most human babies, calves, and colts are born head first. Because of his tadpole shape, large head, stiff neck, and long tapering rear, he has no choice but to back into the outer world. There is no time to spare in his birth and no time to correct a mistake. Only for a moment is the baby whale in danger of being trapped in the birth passage or choked by the five-foot umbilical cord.

As mother and calf roll in the wash the cord snaps. The baby opens a pink mouth with knobby, toothless gums and seems suddenly to smile, for the upturned corners of his mouth break into a satisfied smirk. This is illusion, of course; the smile of a whale is a built-in feature with which it is endowed at birth and retains throughout life. A land creature may yawn and snarl and frown, may screw up its forehead and grow wrinkled in old age; the face of the whale stays round and firm — expressionless but for the rolling of the eyes and the clapping of the jows.

In body form the Little Calf is shapely like his mother, though his tail flukes are pale and crumpled and folded at the corners. While in the womb he had lain like a bent bow, his back arched. Now he is straight and supple. He is almost black, with splotches of white and gray on his slick, rubbery body.

Most whales are in fact dressed in monotonous patterns of black, gray, and white, relieved here and there with faint washes of pink or yellowish. They have no need for brighter coloring, since, as far as anyone knows, they have poor color vision.

In the meantime, off to the north, twenty pilot whales come roistering in with soft explosions, all abreast like troopers. Their foaming wake and the *chuff-chuff-chuff* of their breathing startle the Little Calf. He presses to his mother's side. She makes no move, and the school tumbles on toward the horizon. Had these been killer whales, though, she would have sensed the special underwater sounds of the raiders a mile or more away and would have been ready for battle.

The members of the family of the Little Calf are cruising in a side circle which will bring them back in a week to the same spot. Here and there a whale rests at the surface with full stomach, while others move ahead. A triangle tail rises high in the air for a moment as a whale gains momentum for his long descent, or "sound." The placid water roils at the scene of a petty skirmish between two jealous mothers, each with newborn calf. In the still air of the afternoon the little sounds are few and far between, like the whisperings of a desert land, though the sea below is all aquiver with subdued noise—the ultrasounds of a thousand whales communicating with one another and holding their group together by invisible cords. The whales forge slowly ahead, feeding and loafing and playing, quarreling a bit but leading in general a quiet life. The time of excitement is six full months away — the season of courtship.

ON his first day of life the Little Calf blunders round and round his mother's body, seeking blindly for the place he finally finds: a soft spot that tastes good when he nuzzles it. Each of the mother's two nipples is hidden in a deep slit, one on each side of the belly, far behind the navel. He works his little mouth into position and presses hard; his mother responds by extruding one nipple at a time and squirting a strong flow of milk against the back of his throat. Occasionally the mother turns on her side, with her breast nearly out of the water while the calf, lying parallel with his head in the same direction, holds the teat sideways in the angle of his jaw with his snout protruding from the surface. Such an arrangement seems very awkward, for the Little Calf has no real lips and his lower jaw is narrow, underslung, and twenty inches long.

In his first struggle to nurse, the Little Calf is hindered by a grown female playing the role of "auntie." She lost her own calf by miscarriage in July. With the keen sense of group solidarity of sperm whales, she senses that something is wrong. She glides up to the pair and time and again places her long body between mother and calf. Each time, she is repulsed with a blow from the mother's flukes. She finally moves to a safe distance and listens to the submarine click-talk between the pair, the tones of the mother strong and steady, those of the baby weak and experimental.

For two years the Little Calf will continue to suck at his mother's belly, rising and dipping near the surface of the sea in the long Pacific swell. He follows his mother like a shadow and grows rapidly on his diet of thick milk, over one-third of it pure fat. (The blue dairy milk delivered to my doorstep each morning contains only four percent fat.) His coat of blubber is only an inch thick at first, and during the first few weeks he shivers in the chill of the sea. He will grow steadily throughout the long suckling period, gaining an average of seven pounds a day. Many years hence his blubber coat will have developed into a great firm blanket more than foot thick.

By the end of September the Little Calf is able to match the speed of his family. When a Panama freighter comes throbbing near, alarming the group, he leaps actively beside his mother, spouting freely, keeping up well with the rapid pace of the retreating party. One of the half-grown males bursts clear of the water and floats for an instant against the dark blue sky before he twists in midair and crashes on his side in a glorious fountain of white. Another catches the spirit and breaches four times in a row; he finally coasts along near the surface, tired but content.

But the Little Calf does not fare so well one morning in late September when the sky turns black and sullen, the wind rises, and the fringes of the waves begin to tear away in ragged sheets. The chatter of his family is lost to him in the scream of the gale. One moment beside his mother, at the next he is lifted in a smother of foam, then down a dark, dizzy, sliding trough deep into the sea before he can gasp for breath or clear his nose. His mother senses the trouble and makes a lee for him with her great, comforting back. He spouts air and water for several minutes, then breathes more freely, falls in with the smashing rhythm of the seas, and feels secure. All that day and night his mother stays with him.

Ordinarily, she would have made a dozen or more deep dives during the night to feast on fish and squid. This she must do to replenish her milk supply. It is no hardship, though, for her to fast for a night, for like all whales she can go for weeks without food, living on the reserve fat in her body. Not content perhaps, she is accepting, suffused with what we humans call mother love. Like many a denizen of the open sea she has ridden out a hundred storms, wherein to feed or try to feed would be more painful than simply to wait.

The sky clears in the new dawn. The members of the family convene and begin to doze or play under the warm rays of the sun. A great sea, two hundred feet from crest to crest, is running as slick and silent as oil.

—Victor B. Scheffer (1969)
Year of the Whale

Whale Watchers: Charles M. Scammon (1860)

In October and November the California Grays appear off the coast of Oregon and Upper California, on their way back [from the Arctic] to their tropical haunts, making a quick, low spout at long intervals; showing themselves but very little until they reach the smooth lagoons of the lower coast, where, if not disturbed, they gather in large numbers, passing and repassing into and out of the estuaries, or slowly raising their colossal forms midway above the surface, falling over on their sides as if by accident, and dashing the water into foam and spray about them...

...About the shoals at the mouth of one of the lagoons, in 1860, we saw large numbers of the monsters. It was at the low stage of the tide, and the shoal places were plainly marked by the constantly foaming breakers. To our surprise we saw many of the whales going through the surf where the depth of water was barely sufficient to float them. We could discern in many places, by the white sand that came to the surface, that they must be near or touching the bottom. One in particular lay for a half-hour in the breakers, playing, as seals often do in a heavy surf; turning from side to side with half-extended fins, and moved apparently by the heavy ground-swell which was breaking; at times making a playful spring with its bending flukes, throwing its body clear of the water, coming down with a heavy splash, then making two or three spouts, and again settling under water; perhaps the next moment its head would appear, and with the heavy swell the animal would roll over in a listless manner, to all appearance enjoying the sport intensely.

The World Below the Brine

Walt Whitman

The world below the brine,
Forests at the bottom of the sea, the branches and leaves,
Sea-lettuce, vast lichens, strange flowers and seeds,
the thick tangle, openings, and pink turf,
Different colors, pale gray and green, purple, white,
and gold, the play of light through the water,
Dumb swimmers there among the rocks, coral, gluten,
grass, rushes, and the aliment of the swimmers,
Sluggish existences grazing there suspended, or slowly
crawling close to the bottom,
The sperm whale at the surface blowing air and spray,
or disporting with his flukes,
The leaden-eye shark, the walrus, the turtle, the
hairy sea-leopard, and the sting-ray,
Passions there, wars, pursuits, tribes, sight in those
ocean-depths, breathing that thick-breathing air, as so many do,
The change thence to the sight here, and to the subtle air
breathed by beings like us who walk this sphere,
The change onward from ours to that of
beings who walk other spheres.

—Walt Whitman

At Home With Right Whales

Roger S. Payne, Ph.D.

Every armchair whale-watcher's dream is to actually watch the great whales at home, undisturbed in their natural environment. Since 1971, zoologist Roger Payne has been living this dream, as he, his wife Katy, and their four children have lived in an out-of-the-way research station at Peninsula Valdes in Patagonia, on the remote coast of Argentina. Following up his research on humpback whales and their songs, Payne began a study of the southern right whales—their anatomy, behavior, acoustics, and population changes. (The study was sponsored by the New York Zoological Society and the National Geographic Society.)

. . . One of the things that convinced me right whales were ideal for study is that, alone among all whales, they are adorned with series of peculiar growths called callosities. On every right whale the number, size, shape and placement of callosities are unique, making it possible for us—and presumably the whales too—to tell individuals apart on sight.

Callosities consist of thickened white skin, sometimes many inches deep, with a rough outer surface. . . . The largest callosity appears as if perched on the whale's head. The old whalers called it the "bonnet," and indeed from some angles it does look somewhat like a 19th-century lady's headgear. . . .

In our studies of callosities we discovered an interesting parallel with human beings. Callosities have scattered hairs growing from them. Curiously, right whales' facial hair grows in the same places that a human's does, and only in those places. The whales appear to have what we call mustaches, as well as eyebrows, beards and even sideburns!

Callosities seem to serve several functions. One of them is to act as a sort of splash deflector, preventing water from entering the whale's blowhole. Other species of whale also have splash-deflecting structures, but these are usually fairings, or ridges, around the blowhole, quite different in character from the callosities of right whales.

In aggressive situations callosities may have another use. Right whales sometimes rub their heads across a competitor's flank or back. Since a whale's skin is very soft, even the slightest brush from a crusty callosity could be painful.

Several of our right whales have white spots on their backs that, in addition to

---

Excerpts from "At Home with Right Whales" by Roger S. Payne in *National Geographic,* March 1976; reprinted by courtesy of National Geographic magazine.

callosities help us keep track of individuals. Y-Spot—or Adele, as we later called her—had a calf in 1971 and then vanished for two years. I thought she had died, but she returned in 1974 with a new calf. To us it was a wonderful reunion. It also taught us that the adult's callosity patterns are constant over long periods and that some females may breed only once every three years.

Such a low birthrate would help to explain the very slow recovery of one of the rarest whale species from two centuries or more of intensive hunting. From a population running to tens of thousands in the past, the southern right whale today numbers perhaps fewer than 1,500 individuals. . . .

One form of behavior among right whales looks similar to mating maneuvers but is actually a fascinating form of play. Of all the things we have so far learned, this one delights me most: Right whales appear to sail! As far as we know, they are the only marine creatures, other than some jellyfish, that use the wind for propulsion. Right whales, however, do it as a game.

When a whale sails, it usually does so alone. The tail is "set" straight up above the surface, and at right angles to the wind. Sailing is usually associated with other, more obvious forms of play. When a whale is carried by the wind into shallow water so that its head is bumping along the bottom, it will return upwind, circling around for another sail, like an otter back up the bank for another slide. As the whale does this, it rollicks and galumphs along in a variety of games we have frequently observed. Sailing is done in winds of from five to thirty knots, and can go on for three or four hours.

When it comes to this sport, young whales frequently have a lot of trouble attaining the proper set of their tails. What starts as an apparent effort to sail can degenerate into a wild lobtailing session—that is, with the calf throwing its tail into the air and bringing it back down with a thunderous slap.

At other times lobtailing seems to have a definite function. As with flipper-slapping and breaching, it makes a resounding noise. Though we had looked on such behavior as playfulness, we arrived at a second explanation for the noisy surface displays after systematically recording their occurrence in the bay.

Lobtailing, we found, begets lobtailing. When one whale starts lobtailing, or slippering, or breaching, another nearby will follow suit. Thus, in the most elementary sense, the actions are a form of communication. . . .

—*Roger S. Payne, Ph. D.*

3 Killer Whale at Sea World.
(Photo courtesy of Randall R. Reeves)

Old-time whalemen already knew that whales played and rollicked and cavorted. . .

Oh! jolly is the gale,
And a joker is the whale,
A' flourishin' his tail,—
Such a funny, sporty, gamy, jesty, joky, hoky-poky lad
Is the Ocean, oh!

The scud all a flyin',
That's flip only foamin';
When he stirs in the spicin',—
Such a funny, sporty, gamy, jesty, joky, hoky-poky lad
Is the Ocean, oh!

Thunder splits the ships,
But he only smacks his lips,
A tastin' of this flip—
Such a funny, sporty, gamy, jesty, joky, hoky-poky lad
Is the Ocean, oh!

—Whaling song

The Singing Whale

Roger S. Payne, Ph.D.

Whales are singularly aural creatures, relying on sonar, a variety of cries and sounds, and acute hearing for communication across vast ocean distances. But unique (apparently) even among whales are the clearly organized "sonic repertoires" of the humpback whales, which, like bird songs, include themes that occur in repeated sequences. The best-known researchers into the songs of the humpback whale are Roger and Katy Payne; he is a research zoologist for the New York Zoological Society, and both are trained musicians. Dr. Payne's early analyses used recordings of whale sounds made in Bermuda by Frank Watlington. Spectrograms of these recordings were analyzed musically by Payne and Scott McVay in an article in Science *in August 1971. They found that untis and themes combined into the song of an individual whale, and discovered that "except for the precise configuration of some units and the number of phrases in a theme, there is relatively little variation in successive renditions of any individual humpback's songs." There were several song types around which individual whales built their songs, as well as a very general song pattern common to the whole species.*

After the first commercial recording, Songs of the Humpback Whale, *these mysterious songs intrigued the public and many singers and composers. By the time of the second record,* Deep Voices, *the Paynes had discovered even more—for instance, that all whales in a herd sing the same song each year and then change to a different song the following year.*

Reading about whales singing is not really a substitute for hearing them (just as hearing them on a recording is probably not a substitute for hearing them through a hydrophone on a soft, starry Bermuda night) but, nevertheless, Payne's descriptions make these songs immediate, real and moving.

There are three humpback whales singing at various times in the selection "Three Whale Trip." There is also a lot of ocean noise. The winds had been strong the day before this recording was made and during the day the sea was still running high. Both of the *Twilight's* [the Paynes' sailboat] hydrophones were located near the ocean surface, where wave noise is loudest. After a few moments of listening, however, you will learn to hear much as a whale probably does, ignoring the background noises and focusing on the whale songs.

Dr. Payne made this recording from a sailboat near Bermuda during his studies of the Humpbacks there. "We found one spot,"

The Singing Whale
Quoted from the liner notes for *Songs of the Humpback Whale* and *Deep Voices,* by courtesy of the New York Zoological Society and Roger S. Payne.

says Dr. Payne, "where the sounds of whales blended in a very lovely way. We occasionally stopped there to listen while on our way to make recordings at other locations. The 'Three Whale Trip' was recorded at that favorite listening spot on an occasion when the whales sang all day and all night. . . . As you listen to this recording, I wish only that it could convey to you the pleasant circumstances under which we made it. Through the whole night we listened to the whales, taking turns at the headphones in the cockpit, lulled by the smooth rolling of the boat. Far from land, with a faint breeze and a full moon, we heard these lovely sounds pouring out of the sea."

The title of this selection ["Three Whale Trip"] refers to more than the day's trip on which the recording was made or the voyage of the three whales who paused off Bermuda to sing that day. By playing on the contemporary meaning of the word "trip" as a mental voyage, this title acknowledges what has been discovered time and again by people who have listened to the whales sing: the songs produce an extraordinary inner experience for anyone who lets them into his mind.

* * *

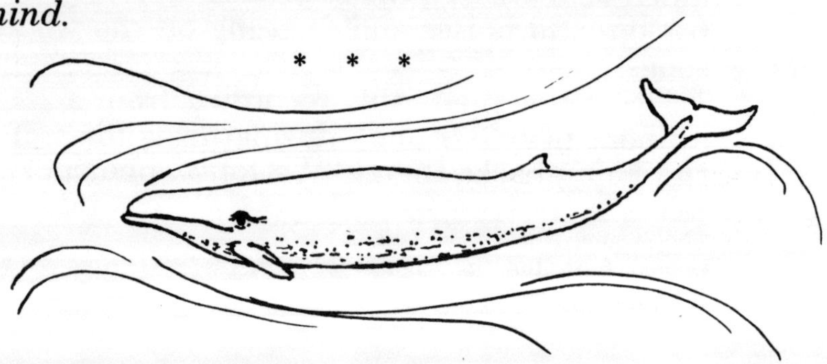

On Deep Voices, *the second whale songs record, Roger Payne describes the mood and settings for several of the recordings, and gives some fascinating observations of the whales themselves.*

Sometimes humpback whales form small, tight groups like packs of dogs that move fast at the surface, breathing nearly in synchrony and with much cavorting and social interaction. When they encounter a boat they stop and approach cautiously and then silently withdraw, milling about for several minutes. At such times they seem not to know what to do and keep close together, pressing against one another's bodies. On a few such occasions these groups have become bold and ended their stay with us by making a mock charge at the boat. This was such an occasion. They grouped about 100 yards away trumpeting and grunting alternately, then turned and plowed towards us howling loudly, and coming so fast and unswervingly we were convinced we were going to be rammed and sunk. Only at the last possible instant did they change direction and dive deftly under the boat, passing beneath us in complete silence but so close to the hydrophone you can hear the gurgling water from their wakes. . . . The winds around Bermuda often blow for long periods and without let. At such times the seas build and build until everything is rolling and reeling and toppling, and all one can do is hang on and wish it would quit. Sometimes the wind dips abruptly and when it does there's no press of air on the sails to hold the boat steady, and so it pirouettes and bows and curtsies clumsily and

C. M. Scammon del. HUMPBACKS LOBTAILING BOLTING BREACHING AND FINNING. *Britton & Rey Lith*

4
***Humpbacks lobtailing, bolting, breaching and finning. Engraving courtesy of Scammon's* Marine Mammals of the North-West Coast of North America.**

heavily as though drunk. It's very uncomfortable aboard. The old sailors call this a "leftover sea running."

We made this recording on such a day in April. We had been out all through a very rough night, and by morning were very tired, and about ready to return when the wind quit all at once. I put down the hydrophones right away and started listening. The contrast was amazing. Beneath all that wildly gyrating surface whitened by bursting waves, undulating and twisting and heaving about, were the sweetest, serenest, loveliest sounds one might ever hear. Clearly the whale cared not one fig for the moods of the ocean and had its own mood which transcended whatever chaos of motion the weather might throw at it. You can hear the whale clearly in this recording, but you may barely notice the sea.

. . . Every spring as if from nowhere the humpback whales mysteriously appear around Bermuda. During the three or four months of their presence the ocean is never silent. It is filled with their songs. But they also have another vocabulary which includes sounds that we never hear in songs. These are social sounds, herd noises that are made in little bursts and followed by long silences. They are often frantic and close together as though there was some momentary squabble in their otherwise serene and gliding lives.

. . . When we first began studying humpback whales, we didn't realize that they sing all through the night. The first time we heard a night chorus it was an accident. We had been drifting far off shore all afternoon and the weather being fair we decided to stay out overnight. We were out of sight of land over Argus bank, a shallow plateau that rises from deep water about 30 miles from Bermuda—a place where humpbacks congregate. Whenever they sang in deep water suspended far above the abyss, the echoes were distinct and one got the sense of vast, vaulted spaces as if the ocean were a great cathedral, but now we could tell that they were up on the bank as there were no echoes, and their voices sounded flat and lackluster.

As darkness fell, the wind died and we were carried only by the slow ocean currents. We passed that night immersed in the sound of whales. It was a strange and lovely sound as if of sleepers singing, a mingled, mazy drone of voices. It was as if they sang as they drifted off to sleep right through from consciousness to unconsciousness. . . .

. . . When Scott McVay and I became aware that humpback whales sing songs, the first thing we both wanted to see was whether other species did too. Scott went north to Alaska to record bowhead whales while I went to the southern hemisphere to learn about right whales. I had found an area where right whales live so close to shore we could study them without the major expense of ocean-going vessels. It was Peninsula Valdez in Argentina, an enormous cape enclosing two bays, about half way between Buenos Aires and Tierra del Fuego on the windy Patagonian coast. In this region for five months of each year, a small herd of right whales (the rarest of the large whale species) is resident. At first we lived in tents but later built a field station with modest living quarters to supplement the tents. We named our corner of San Jose Bay (the northernmost bay on Peninsula Valdez) Camp Bay. A cove to the west where whales often congregated became Whale Bay. . . .

For most of five years my family and I lived on Peninsula Valdez surrounded by a million acres of wilderness, an infinite ocean, but in the midst of a tiny band of right whales. We installed hydrophones on the sea bottom near camp and led the wires ashore to where we had a loudspeaker that could play all day and bring the voices of the whales into our lives. Sometimes we heard distant cries, haunting and all but lost against the hiss of waves out in the bay. At others the sounds were close and immediate as the whales swam and courted in front of camp.

. . . I made this recording of underwater sounds on a nearly calm morning when there were ten whales in front of camp: four mothers with calves, a lone adult and a yearling. As mothers, calves and yearlings don't say much, the lone adult probably made most of these sounds. I suspect it was a male, but I have no idea what the sounds mean. For a long while the loner stayed at the surface right over the hydrophone (once dragging its tail right across it) and then we could hear the waves lapping against the whale's flanks. This was the first time it occurred to me that on rough days whales must experience a dramatic contrast between the relative calm beneath the ocean surface and the deafening wash and crash of waves against their heads when they surface to blow.

In this cut you can hear breaths from underwater, and a calf slapping the surface far off, answered by one slapping nearby. This seems to be very primitive form of communication in right whales, perhaps a way of finding out where each other is. In the background you can hear a hissing surge. These are pebbles on the beach being dragged back by each retreating wave. The beach in front of camp is composed of tiny polished pebbles of every subtle and vibrant color like a heap of gems. The children often made collections—no two alike—and a single square foot could take all morning to explore properly.

I can remember sitting on the beach with my arm buried to the wrists in pebbles listening to the whales' fantastic sounds coming in through a hydrophone. . . we often heard whinnies, growls, moans, chuckles, chirps, gurgles, snickers, creaks, burps, groans, cries, keens and so on ad infinitum. We found all this variety, but no songs.

A long time later I saw Scott McVay and learned with disappointment that he had not heard songs from bowheads either. It was really a letdown. I had so hoped that at least one other species sang long, haunting songs... but this new mystery. . . of why all that variety. It was an unbelievable array of sounds, almost none of which we understand but at least some of which must surely have meaning for the life of the herd.

. . . Right Whales once filled many of the bays and inlets on earth. They were in the Delaware Bay, the Chesapeake, in New York and Boston harbors, in San Francisco and many bays on the European coast line. Diarists from the *Mayflower* marvelled at the whales they saw as they entered Massachusetts Bay. The surfaces of all these bays were alive with the slow, restful breathing of whales. This is how the shores of the world sounded during at least the first twenty million years of the existence of whales. It is only in the last two hundred years that the scene has changed, so that now children whose greatgrandfathers must have delighted in hearing these sounds cannot even imagine what life was like such a short time ago.

—Roger S. Payne

Whale Watchers: Victor Scheffer

Like humans, some whales live in families, play in the moonlight, talk to one another, and care for one another in time of trouble. We humans can feel a bond there. But is it not enough simply to respect whales as whales? Instead of wondering whether they can reason, or worry, or think ahead as we humans can, perhaps we should be content to admire them for their unique minds and bodies—for their vital architecture that allows them to exist in a forbidding world of water in which our own naked selves could not long survive.

Quoted from "Exploring the Lives of Whales" by Victor Scheffer in *National Geographic,* December 1976.

5
(Photo courtesy of Randall R. Reeves)

. . . Sand-strewn caverns, cool and deep,
Where the winds are all asleep;
Where the spent lights quiver and gleam,
Where the salt weed sways in the stream,
Where the sea-beasts, ranged all round,
Feed in the ooze of their pasture-ground;
Where the sea-snakes coil and twine,
Dry their mail and bask in the brine;
Where great whales come sailing by,
Sail and sail, with unshut eye,
Round the world for ever and aye. . . .

—Matthew Arnold, "The Forsaken Merman"

6
(Photo courtesy of the National Oceanic and Atmospheric Administration)

2

THE SCIENTIFIC WHALE

. . . There is still much to be learned about these wonderful mammals.

—Edwin H. Colbert

The Nature of the Whale

Bartholomaeus Anglicus, DeProprietatibus Rerum

The author, who was a monk in the 13th century, wrote this "compendium of universal knowledge" in Latin. The work was translated about a hundred years later.

When the Whale hungereth sore, he casteth out of his mouth a vapour, that smelleth as the smell of amber; and fish have liking in that smell, and for the odour and smell of that vapour, they go into the Whale's mouth and be so deceived and eaten. Also in this fish earthy matter hath more mastery than watery, and therefore he is soon great and fat; and so in age for greatness of body, on his ridge [backbone] powder and earth is gathered, and so digged together that herbs and small trees and bushes grow thereon; so that that great fish seemeth an island. And if shipmen come unwarily thereby, unneath they scape without peril—for he throweth so much water out of his mouth upon the ship, that he overturneth it sometime or drowneth it. And also he is so huge in quantity that when he is taken, all the country is better for the taking.

Also he loveth his whelps with a wonder love, and leadeth them about in the sea long time. And if it happeth that his whelps be let with heaps of gravel and by default of water, he taketh much water in his mouth and throweth upon them, and delivereth them in that wise out of peril. And he setteth them always between himself and the sun on the more sicher side. And when strong tempest ariseth, while his whelps be tender and young, he swalloweth them up into his womb; and when the tempest is gone and fair weather come, then he casteth them up whole and sound.

Also against the whale fighteth a fish of serpent's kind and is venomous, as a crocodile; and then other fish come to the Whale's tail, and if the Whale be overcome, the other fish die. And if the venomous fish may not overcome the Whale, then he throweth out of his jaws into the water a fumous smell most stinking; and the Whale throweth out of *his* mouth a sweet-smelling smoke, and putteth off the stinking smell, and defendeth and saveth himself and his in that manner wise.

From *De Proprietatibus Rerum*, a "compendium of universal knowledge" written in Latin in the 13th century by the monk Bartholomaeus Anglicus, and translated about one hundred years later.

—Bartholomaeus Anglicus,
DeProprietatibus Rerum

The Nature of the Whale
Reprinted from *The World of Animals,* compiled by Joseph Wood Krutch (Simon & Schuster, 1961). Used by permission of the publisher.

Whale Watchers: Pliny the Elder

. . . The largest number of animals and those of the largest size are in the Indian Sea, among them whales covering three acres each and sharks 100 ells long; in fact, in those regions lobsters grow to six feet long and also eels in the river Ganges to 300 feet. . . .

The largest animals in the Indian Ocean are the shark and the whale; the largest in the Bay of Biscay is the sperm whale (*physeter*) which rears up like a vast pillar higher than a ship's rigging and belches out a sort of flood. . . .

Whales even penetrate into our seas. It is said that they are not seen in the Gulf of Cadiz before midwinter, but during the summer periods hide in a certain calm and spacious inlet and take marvellous delight in breeding there; and that this is known to the killer whale [Pliny says *orca*], a creature that is the enemy of the other species and the appearance of which can be represented by no other description except that of an enormous mass of flesh with savage teeth. . .

Whales have their mouths in their foreheads and consequently when swimming on the surface of the water, they blow clouds of spray into the air. It is universally admitted that a very few other creatures in the sea also breathe—those whose internal organs include a lung, since it is thought that no animal is able to breathe without one. Those who hold this opinion believe that the fishes possessing gills do not alternately expire and inspire air, and that many other classes even lacking gills do not—an opinion which I notice that Aristotle held. . . .

Whales do not possess gills nor do dolphins. These two genera breathe with a tube that passes to the lung, in the case of whales from the forehead and in the case of dolphins from the back. . . .

The swiftest of all animals, not only those of the sea, is the dolphin; it is swifter than a bird and darts faster than a javelin. . . . For a voice they have a moan like that of a human being; their back is arched and their snout turned up, owing to which all of them in a surprising manner answer to the name of "Snubnose" (*Simo*) and like it better than any other.

The dolphin is an animal that is not only friendly to mankind but is also a lover of music, and it can be charmed by singing in harmony, but particularly by the sound of the water organ. . . .

—*Historia naturalis*
(A.D. 77)

From *Historia Naturalis (Natural History)* by Pliny the Elder (Caius Plinius Secundus), translated by H. Rackham.

The Whale's Family Tree

F.D. Omanney

Although there is no fossil record and no certain history of the whales' original land ancestors, speculation is still possible and interesting. . . .

Whales, porpoises and dolphins belong to the mammalian order Cetacea (Greek, *ketos*— a whale). The word "whale" itself, similar in all northern European languages (Norwegian *hval,* German and Dutch *wal*), is closely related to the word "wheel" and refers to the manner in which the animal's back seems to be turning over or revolving in the water. The Cetacea stand, in relation to other orders of mammals, near the even-toed or cloven-hoofed Ungulates (cattle, sheep, goats, pigs and camels—as opposed to the odd-toed, single-hoofed Ungulates such as the horse). The resemblances to the even-toed Ungulates are highly technical, but we may note the small number of young, usually one at a time, the position of the mammary glands and the shape of the penis. Internally there is the structure of the genital organs, the arrangement of the placental membranes before birth, the form of the stomach (not unlike that of the domestic cow) and the chemical structure of the body proteins.

All the Cetacea breathe air and at each exhalation blow out a stream of water vapour called the "spout," for which reason the nostrils or single nostril on top of the head are referred to as "blowholes." In the great whales the spout is a most impressive sight, like the exhaust of a great engine pounding up from the deep. It may be a fine, narrow upstanding jet, as in the great whalebone whales, or it may be quite small and inconspicuous, as in many porpoises and dolphins. In all Cetacea a layer of blubber sheaths the body from head to tail and all the tissues, muscles, bones and viscera are impregnated with oil which helps to insulate them in their cold watery world. Over the flukes and flippers the blubber layer is thin and fibrous and it is thin over the skull and belly, but over the flanks and back it may be as much as a foot thick in whalebone whales. In a large fat animal in good condition after months of feeding in southern seas the blubber layer may be as much as two feet thick on the flanks.

Ancestors. Some hundred million years ago the whales' forbears lost contact with the land and took to their present entirely aquatic life. The great Antarctic whales, indeed, pursue their food, mate and bear their young many hundreds of miles from any land.

The fossil record gives us almost no help in deciding how and where the change from land to sea took place. Fossil creatures of archaic form, yet undoubtedly whales, appear quite suddenly in the record with no antecedents,

The Whale's Family Tree
Reprinted from *Lost Leviathan,* by F.D. Ommaney, by permission of Dodd, Mead & Company.

the oldest remains being about 55 million years old. Possibly the lack of any record prior to this is due to the fact that the change from land to water took place in sandy or swampy estuaries where no fossils could be formed. Having left the land the whales' ancestors probably continued their evolution in the open sea where, again, no fossils could be formed.

The earliest fossil whales known had teeth in their jaws which resembled those of an extinct race of early Carnivores. (Carnivores, as their name implies, are the flesh-eating mammals—dogs, cats, bears, seals, walruses.) Both the Carnivores and the even-toed Ungulates (cattle, etc.) can be traced back to an ancestor common to both but long ago extinct which was something like our modern Insectivores (hedgehogs, moles, shrews). If we find it difficult to imagine a small creature like a shrew as a progenitor, no matter how remote, of the colossal whalebone whales of the Antarctic, we must remember that mere size is no clue to descent. Most races began as small creatures at the outset and only reached great size as they approached the end. We know that the whalebone whales were comparatively recently much smaller than they are now, while the most successful and numerous Cetacea, dolphins and porpoises, are quite small.

In 1845 a certain Dr. Albert Koch dug up some enormous vertebrae and added some others and a skull. He mounted them into a skeleton which he exhibited in New York and in Europe as the first sea serpent, 114 feet long. He called it *Hydrarchos*, the water chief. Richard Owen pointed out that it was really made up of two separate animals and was, in fact, the remains of an early whale. As he reconstructed it the creature was still 55 feet long and had a long, sinuous tail with which it must have been able to swim quickly with a whip-like action. It was a snake-shaped animal with variegated teeth and a skull resembling that of a dog in shape. The fore-limb was a flipper with five fingers in the hand and with an elbow. There were traces of a thigh bone.

Since the discovery of these early serpent-shaped whales many other remains have been found in various parts of the world, especially in Egypt and on the Gulf Coast of the U.S.A. Some of them were shaped like sea serpents and others were like modern porpoises, swimming swiftly with up and down strokes of a flattened tail. Yet none of these can be looked upon as the ancestors of any whales now living.

We can only imagine what the small creature was like which waded out into the waters of the steadily advancing shallow sea typical of the Cretaceous period about a hundred million years ago during which the chalk deposits were laid down.

This ancestral animal was part insectivore and part carnivore. It had a five-fingered foot and hand and an elongated skull like a dog with teeth differentiated into incisors, canines, pre-molars and molars, as are the teeth of carnivores today. It has been suggested that the skin may have been partly, or even wholly, covered with small plates of bone because these are found in a number of fossil whales. It probably lived in estuaries where there were gently sloping beaches of sand or mud. It may have fed on crustaceans or small fish and took to wading after them farther and farther out into the shallows. In time it took to swimming after them more and more quickly, using its tail for propulsion and its fore-limbs as stabilizers. In the course of time its five fingers became bound together into a flipper while its hind limbs, probably held back motionless against the body and taking no part in swimming, slowly atrophied and disappeared, the tail becoming widened out to form flukes. Some of these animals swam with whip-like movements of the tail, but it was from those which swam by means of vertical strokes of the tail that the modern whales came in the fullness of time. Other small animals, part insectivore and part early carnivore, living under different, perhaps drier conditions, took to running after terrestrial prey and became the ancestors of modern Carnivores. Others again became vegetable feeders and gave descent to the Ungulates.

Whalebone and Toothed Whales. The modern order Cetacea is divided into two sub-orders which do not seem to be very closely related to one another. One (Mystacoceti [more usually, Mysticeti], from the Greek *mystax*, a moustache, *ketos*, a whale) comprises those whales which have no teeth in their mouths but have instead whalebone plates, or baleen as it is called. The other (Odontoceti, from the Greek *odontos*, teeth) comprises whales which have teeth in their mouths either in both jaws or in the lower jaw only. In the first sub-order are the great whalebone whales which have been hunted for centuries in the Arctic regions and, since the beginning of the present century, in the Antarctic. Those in the second include the Sperm whale which has been hunted all over the world since the beginning of the eighteenth century but whose pursuit and capture by mankind was probably taking place many centuries before the Christian era. It also includes the Killer whale or Grampus and many other smaller whales, and those we speak of as porpoises and dolphins.

The whalebone or baleen whales (Mystacoceti [Mysticeti]) are filter feeders. Like a great many different kinds of aquatic animals, both vertebrate and invertebrate, they do not feed by means of a mouth armed with teeth but strain out of the water large quantities of minute food too small to be pursued, caught and devoured with teeth. Among invertebrates many crustacea, all bivalve molluscs and sea squirts feed in this manner. Among vertebrates many shoaling fishes, such as herrings and sardines, and the enormous basking shark feed by straining multitudes of small drifing animals and plants out of the water by means of comb-like processes on the gill clefts. Baleen whales filter their planktonic food out of the water by means of a series of triangular flexible plates, made of a horny substance we call whalebone or baleen, hanging down from the upper jaw. These are placed one behind the other with only an inch or so between them and their inner edges are frayed out so as to form a mat of coarse hairs over the roof of the mouth. On this the small animals and fishes on which the whale feeds become entangled. The whale then closes its mouth and, by means of its fleshy tongue, forces the water out between the whalebone plates and over the rampart of the lower jaw. With the mouth closed it then sweeps the food back into the throat, perhaps by means of the tongue, though just how this is done is not certain.

The toothed whales (Odontoceti) are all flesh eaters except one small vegetarian dolphin that lives in the Cameroon River, West Africa. Some, such as the Sperm whale, feed on squids while others, such as dolphins and porpoises, feed mainly on fish. The Grampus or Killer whale devours penguins, seals, fish or the flesh of other whales.

—F.D Ommaney

7
Head-on with a fin whale (or finback), one of the largest baleen, or whalebone, whales. (Photo courtesy of Jean Laruin)

The Return to the Sea

Edwin H. Colbert

Of all mammals the whales and porpoises, or cetaceans, are certainly the most atypical, and in many ways the most highly specialized in the extent to which they have diverged from their primitive eutherian ancestors. These mammals must have had an ancient origin, for no intermediate forms are apparent in the fossil record between the whales and the ancestral Cretaceous placentals. Like the bats, the whales (using this term in a general and inclusive sense) appear suddenly in early Tertiary times, fully adapted by profound modifications of the basic mammalian structure for a highly specialized mode of life. Indeed, the whales are even more isolated with relation to other mammals than the bats; they stand quite alone. Therefore, it seems evident that the whales, having separated from the ancestral eutherians at an early date, enjoyed at the outset a series of extraordinarily rapid evolutionary changes that made them by middle Eocene times completely adapted for life in the ocean.

The whales returned to the sea to imitate the fishes after a fashion and to take the place of the ichthyosaurs, the fishlike marine tetrapods of Mesozoic times. In becoming adapted to a life in the open ocean the whales, as descendants of fully land-living ancestors, were faced with the same problems that had been encountered and overcome millions of years previously by the first ichthyosaurs.

The problems to be surmounted by the ancestral whales in their new environment were in short, those of locomotion, of respiration or breathing, and of reproduction. The adaptations to locomotion in the water involved a streamlining of the body, the development of a fishlike tail as the main propeller, and the transformation of the legs into balancing paddles. Like the ichthyosaurs of the Mesozoic, the whales retained the lung breathing that was their heritage from land-living ancestors, and in so doing they evolved adaptations that increased the efficiency of respiration. Of course embryonic development in whales posed no particular problems for these mammals, but there were special adaptations for the survival of the young in the water, from the moment of birth on.

The return of whales to the sea is a fine example of convergence in evolution. In following this evolutionary trend the whales have shown many adaptations that are remarkably similar to those of the ichthyosaurs, yet the ancestors of these two groups of tetrapods were quite distinct—cotylosaurian reptiles on the one hand and primitive eutherian mammals on the other. Convergence such as this illustrates the remarkably similar adaptations by dissimilar

---

The Return to the Sea

animals to an environment that imposes stringent limitations upon its inhabitants.

Whales as Marine Vertebrates. As already mentioned, whales are fishlike vertebrates, highly streamlined for efficient swimming. The body is torpedo shaped, and there is no visible neck distinct from the trunk. There is no hair on the body of the modern whales, and it is likely that this adaptation developed early in the history of these mammals. The smooth skin makes a sleek surface that offers little resistance to the water as the animal moves forward. Since whales are warm blooded, they have developed a heavy layer of fat or blubber beneath the skin, to give the necessary insulation that in most mammals is furnished by the covering of hair.

The vertebrae are much alike and numerous, making for a flexible backbone, but the neck vertebrae are commonly shortened and fused into a single bony mass. Long muscles and tendons attach to the vertebrae and run back to the tail, to furnish the propulsive force that drives the animal through the water. The whales are unique among marine vertebrates in that the tail terminates in a horizontal fin that moves up and down, rather than in a vertical fin that moves from side to side. The horizontal tail fin of the whale, generally called flukes, is a neomorphic structure in these mammals and, although stiff and strong, contains no bones. Likewise the whales frequently have a fleshy dorsal fin, another neomorph or new structure. This fin is a stabilizer that prevents the animal from rolling, and may be compared in function with the fleshy dorsal fin of the ichthyosaurs or the bony dorsal fins of fishes.

8 Finback whales

As in the ichthyosaurs, the limbs in whales have been modified as paddles. However, in all known whales, both fossil and recent, the pelvis and hind limbs are reduced to mere vestiges; only the front limbs remain as functional paddles. The arm bones are short and flattened, the wrist bones are flattened discs, and the fingers are commonly greatly elongated by multiplication of the phalanges as supports for the paddle.

In all but the earliest whales the external nostrils are shifted to the top and the back of the skull, to form the "blowhole" so characteristic of these mammals. The nostrils can be closed by valves, and the lungs are highly elastic and extensible, for taking in great quantities of air. Whales are able to remain submerged for long periods of time—as long as an hour in some of the large whales—and some of them can dive to great depths. In line with thse remarkable accomplishments, there are profound adaptations in the physiology of the whales.

In addition to the modifications in the skull

brought about by special adaptations for breathing, the whales show marked specializations of the ear. The tube to the external ear and the eardrum are drastically reduced, and it is obvious that whales do not hear sounds in the same manner as other mammals. Instead they are very sensitive to vibrations in the water, which are transmitted to a heavy, shell-like bone formed by a fusion of the periotic bone and the auditory bulla, and separate from the rest of the skull. Indeed, modern experimental work has shown that the cetaceans, like the bats, have a very sophisticated sonar apparatus by means of which they can locate and identify objects in the water. Moreover, whales and porpoises communicate with each other by projecting underwater "sounds" or vibrations of varying frequencies; their system of communication seems to have reached a rather high stage of development. In contrast to this highly developed sense, whales have no olfactory sense. Consequently olfactory lobes are lacking, and the brain is large and rounded. The brain is highly developed in these mammals, and recent experiments indicate that whales are remarkably intelligent. The brain and the sensitive auditory apparatus are telescoped into a small space at the back of the skull; the remainder of the skull is made up of very long jaws.

Whales are and have been throughout their history carnivorous, living on large invertebrates, fishes, and even other whales, or on microscopic marine animals known as plankton. In the first category are the toothed whales, ancient and modern. Those in the second category are the whalebone whales, in which the teeth are suppressed and transverse plates of fibrous keratin or baleen hang from the roof of the mouth to strain plankton from the water.

As for reproduction, young whales are very large and well formed when they are born. As soon as they are born, the mother pushes them to the surface so that they can get their first breaths of air. From that time on the baby is able to swim along with its mother. The mammary glands are enclosed in a sort of pocket, so that the young whale can nurse without shipping a lot of sea water.

Early Whales. Whales appear in the geologic record in sediments of middle Eocene age. These first whales, known as archaeocetes, are characterized by the genera *Protocetus* and *Eocetus* of middle Eocene age and by *Basilosaurus* commonly designated as *Zeuglodon* of upper Eocene age. The early whales were large, and in this respect they indicate an evolutionary trend typical of the whales. Many of these mammals, freed from many of the limiting effects of gravity, have become the giants of the animal world past or present, and even *Basilosaurus* or *Zeuglodon* of the Eocene epoch was some sixty feet in length. In this whale the tail was very long, the forelimbs were modified into paddles, and the hind limbs were suppressed.

The skull in the archaeocetes was somewhat more primitive than the skull in later whales. The bones of the facial region were not telescoped into the back of the skull, as they were in the more advanced whales. There were forty-four teeth, as in primitive placentals, of which the incisors and the canines were simple, sharp cones. The cheek teeth were cusped, but with the cusps arranged in a single fore and aft line, and coming to a high point in the middle. Teeth of this form are seen in other mammals, especially certain seals that live on fish. The nostrils in the archaeocetes were placed in a forward position, not on top of the skull as in the later whales. In general this skull appears as if it might have been derived from a creodont type, but there is little beyond certain general resemblances to support such a relationship. It must be emphasized that the ancestry of the Cetacea is completely unknown at the present time.

Adaptive Radiation of the Modern Whales. During late Eocene or Oligocene times the modern whales arose as descendants of the archaeocetes, and by Miocene times almost all the families of modern whales had appeared. Two lines of cetacean evolution developed from the archaeocete stem. One of these was the toothed whales or odontocetes; the other was the whalebone whales or mysticetes.

Most of the modern whales are odontocetes. In late Oligocene times there appeared some comparatively small odontocetes, designated as the squalodonts, much like the modern porpoises in general appearance, but charactereized by cusped, triangular cheek teeth somewhat similar to the teeth in the early archaeocetes. Evidently the squalodonts, of which the Miocene genus *Prosqualodon* is typical, were forms intermediate between the archaeocetes and

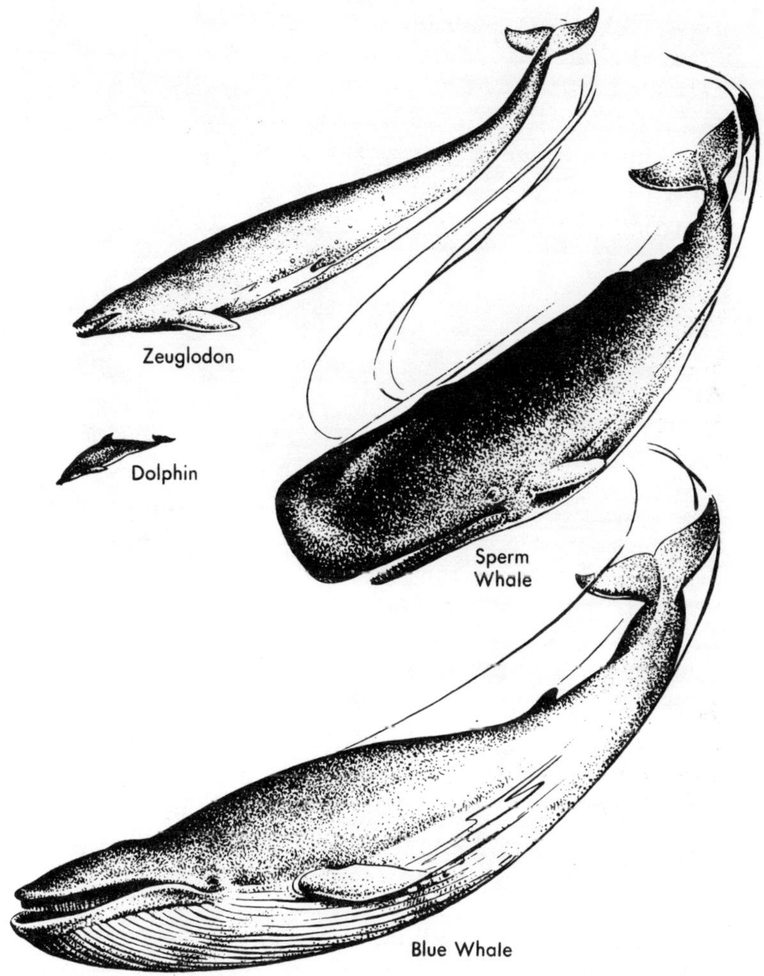

9
A prehistoric whale or archaeocete (Zeuglodon) from the Eocene epoch, about 50 million years ago, is shown in scale with more recent cetaceans—a dolphin and a sperm whale, both toothed whales, and a great blue whale, a baleen whale, about 90 feet in length. (Illustration courtesy of Lois M. Darling)

10
Killer whale at Sea World, San Diego. (Photo courtesy of Randall R. Reeves)

the modern types of whales. In spite of their archaic teeth, the skull was highly advanced, with the nostrils completely dorsal in position and the skull bones modified accordingly. The squalodonts were important whales during the Miocene epoch, but they did not survive long after the beginning of the Pliocene epoch.

Their place in Pliocene, Pleistocene, and Recent times was taken over by the small, toothed whales with which we are familiar, especially the porpoises and dolphins, collectively known as delphinids. These are compact, very swiftly swimming cetaceans that feed upon fish. The teeth are greatly multiplied in number and have the form of simple spikes. A common porpoise of Miocene times was *Kentriodon*; two of the most widely spread of the modern porpoises are *Delphinus*, the common dolphin, and *Phocaena*, the common porpoise of worldwide distribution. Related to these cetaceans are the fierce killer whales, the so-called blackfish, and some of the river dolphins. Other small to medium-sized odontocetes are the narwhals, the beaked whales, and the platanistids that live in the Amazon and the Ganges rivers.

There was an early trend toward giantism in the toothed whales, reaching its culmination in the physeterids or sperm whales. The modern sperm whale or cachalot (*Physeter*) is a giant odontocete with a great, square snout filled with gallons of sperm oil. Peglike teeth are present on the lower jaw, and this whale feeds upon squids. In the great days of whaling the sperm whale was much sought for its oil, and incessant hunting has reduced it from its original vast numbers to the limited herds of the present day. Moby Dick was a white sperm whale.

The largest of the modern whales are the mysticetes or whalebone whales, of much lesser variety than the toothed whales, but none the less highly successful cetaceans. As mentioned above, these whales feed upon plankton, and it may have been the abundance of their food supply that led to the strong trend to giantism among these largest of all vertebrates. The primitive mysticetes were the cetotheres, in which the teeth had been lost and the skull had progressed toward the high degree of modification that is so characteristic of the modern whalebone whales. *Mesocetus*, of Miocene age, was typical of this group.

In the latter portion of Cenozoic times the evolution of the whalebone whales reached its ultimate stage with the development of the skull into a highly arched structure for bearing large plates of baleen, so that the eyes and braincase were limited to a very small posterior region. Some modern representatives of this evolutionary trend are the Greenland whales, the right whales (*Balaena*), the finbacks or rorquals (*Balaenoptera*), and the titanic blue whales. These last may reach lengths of one hundred feet and weights of one hundred and fifty tons. We are wont to look back at some of the giant dinosaurs with awe, yet such modern whales as the blue whale far exceed in size the largest dinosaurs. These are the ultimate extreme of giantism in the evolution of animals.

Whales and Man. A few hundred years ago whales roamed the seas in vast herds. Then there was a demand for whale oil, and the large-scale systematic hunting of whales began, reaching a high point in the last and present centuries. Many species of large whales have been drastically reduced in numbers. Whale oil is not used for lamps as it once was, but the hunting of whales continues on a vast scale, for the oil is used for soaps, the meat is processed for food, and the bone is ground up for fertilizer. Factory ships go out to hunt whales with airplanes, radar, modern whale guns, and other refined apparatus. Consequently many whales are in danger of extermination, and it is only through strict adherence to international treaties that these large mammals can be preserved for posterity. It is to be hoped that they will not disappear from the seas, for the giant whales together with their smaller cetacean cousins are marvelous animals that should never be allowed to vanish at the hand of man. There is still much to be learned about these wonderful mammals.

—Edwin H. Colbert,
Curator
Emeritus of
Vertebrate
Paleontology,
American Museum
of Natural History

The Whale

—Warm and buoyant in his oily mail
Gambols on seas of ice the unwieldy whale,
Wide waving fins round floating islands urge
His bulk gigantic through the troubled surge.
O'er the white wave he lifts his nostril bare,
And spouts transparent columns into air;
The silvery arches catch the setting beams,
And transient rainbows tremble o'er the
streams.

—*Erasmus Darwin*

11 Whale on ice. (Photo courtesy of U.S. Navy)

Whale Watchers: Thor Heyerdahl

I have seen whales in the distance from boats, and I have seen them stuffed in museums, but I have never felt toward the gigantic carcass as one usually feels toward proper warm-blooded animals, for example a horse or an elephant. Biologically, indeed, I had accepted the whale as a genuine mammal, but in its essence it was to all intents and purposes a large cold fish. We had a different impression when the great whales came rushing toward us, close to the side of the raft.

One day, when we were sitting as usual on the edge of the raft having a meal, so close to the water that we had only to lean back to wash out our mugs, we started when suddenly something behind us blew hard like a swimming horse and a big whale came up and stared at us, so close that we saw a shine like a polished shoe down through its blowhole. It was so unusual to hear real breathing out at sea, where all living creatures wriggle silently about without lungs and quiver their gills, that we really had a warm feeling for our old distant cousin the whale, who like us had strayed so far out to sea. Instead of the cold, toadlike whale shark, which had not even the sense to stick up its nose for a breath of fresh air, here we had a visit from something which recalled a well-fed jovial hippopotamus in a zoological gardens and which actually breathed—that made a most pleasant impression on me—before it sank into the sea again and disappeared.

—*Kon-Tiki*

12
(Photo courtesy of U.S. Navy)

The Tail

Herman Melville

Other poets have warbled the praises of the soft eye of the antelope, and the lovely plumage of the bird that never alights; less celestial, I celebrate a tail.

Reckoning the largest sized Sperm Whale's tail to begin at that point of the trunk where it tapers to about the girth of a man, it comprises upon its upper surface alone, an area of at least fifty square feet. The compact round body of its root expands into two broad, firm, flat palms or flukes, gradually shoaling away to less than an inch in thickness. At the crotch or junction, these flukes slightly overlap, then sideways recede from each other like wings, leaving a wide vacancy between. In no living thing are the lines of beauty more exquisitely defined than in the crescentic borders of these flukes. At its utmost expansion in the full grown whale, the tail will considerably exceed twenty feet across.

The entire member seems a dense webbed bed of welded sinews; but cut into it, and you find that three distinct strata compose it:—upper middle, and lower. The fibres in the upper and lower layers, are long and horizontal; those of the middle one, very short, and running crosswise between the outside layers. This triune structure, as much as anything else, imparts power to the tail. To the student of old Roman walls, the middle layer will furnish a curious parallel to the thin course of tiles always alternating with the stone in those wonderful relics of the antique, and which undoubtedly contribute so much to the great strength of the masonry.

But as if this vast local power in the tendinous tail were not enough, the whole bulk of the Leviathan is knit over with a warp and woof of muscular fibres and filaments, which passing on either side the loins and running down into the flukes, insensibly blend with them, and largely contribute to their might; so that in the tail the confluent measureless force of the whole whale seems concentrated to a point. Could annihilation occur to matter, this were the thing to do it.

Nor does this—its amazing strength—at all tend to cripple the graceful flexion of its motion; where infantileness of ease undulates through a Titanism of power. On the contrary, those motions derive their most appalling beauty from it. Real strength never impairs beauty or harmony, but it often bestows it; and in everything imposingly beautiful, strength has much to do with the magic. Take away the tied tendons that all over seem bursting from the marble in the carved Hercules, and its charm would be gone. As devout Eckerman lifted the linen sheet from the naked corpse of Goethe, he was overwhelmed with the massive chest of the man, that seemed as a Roman triumphal arch. When Angelo paints even God the Father in human form, mark what robustness is there. And whatever they may reveal of the divine love in the Son, the soft, curled, hermaphroditical Italian pictures, in which his idea has been most successfully embodied; these pictures, so destitute as they are of all brawniness, hint nothing of any power, but the mere negative, feminine one of submission and endurance, which on all hands it is conceded, form the peculiar practical virtues of his teachings.

Such is the subtle elasticity of the organ I treat of, that whether wielded in sport, or in earnest, or in anger, whatever be the mood it be in, its flexions are invariably marked by exceeding grace. Therein no fairy's arm can transcend it.

Five great motions are peculiar to it. First, when used as a fin for progression; Second, when used as a mace in bottle; Third, in sweeping; Fourth, in lobtailing; Fifth, in peaking flukes.

First: Being horizontal in its position, the Leviathan's tail acts in a different manner from the tails of all other sea creatures. It never wriggles. In man or fish, wriggling is a sign of inferiority. To the whale, his tail is the sole means of propulsion. Scroll-wise coiled forwards beneath the body, and then rapidly sprung backwards, it is this which gives that singular darting, leaping motion to the monster when furiously swimming. His side-fins only serve to steer by.

Second: It is a little significant, that while one Sperm Whale fights another Sperm Whale only with his head and jaw, nevertheless, in his conflicts with man, he chiefly and contemptuously uses his tail. In striking at a boat, he swiftly curves away his flukes from it, and the blow is inflicted only by the recoil. If it be made in the unobstructed air, especially if it descend to its mark, the stroke is then simply irresistible. No ribs of man or boat can withstand it. Your only salvation lies in eluding it; but if it comes sideways through the opposing water, then partly owing to the light buoyancy of the whale-boat, and the elasticity of its materials, a cracked rib or a dashed plank or two, a sort of stitch in the side, is generally the most serious result. These submerged side blows are so often received in the fishery, that they are accounted mere child's play. Some one strips off a frock, and the hole is stopped.

Third: I cannot demonstrate it, but it seems to me that in the whale the sense of touch is concentrated in the tail; for in this respect there is a delicacy in it equalled only by the daintiness of the elephant's trunk. This delicacy is chiefly evinced in the action of sweeping, when in maidenly gentleness the whale with a certain soft slowness moves his immense flukes from side to side upon the surface of the sea; and if he feel but a sailor's whisker, woe to that sailor, whiskers and all. What tenderness there is in that preliminary touch! Had this tail any prehensile power, I should straightway bethink me of Darmonodes' elephant that so frequented the flower-market, and with low salutations presented nosegays to damsels, and then caressed their zones. On more accounts than one, a pity it is that the whale does not possess this prehensile virtue in his tail; for I have heard of yet another elephant, that when wounded in the fight, curved round his trunk and extracted the dart.

Fourth: Stealing unawares upon the whale in the fancied security of the middle of solitary seas, you find him unbent from the vast corpulence of his dignity, and kitten-like, he plays on the ocean as if it were a hearth. But still you see his power in his play. The broad palms of his tails are flirted high into the air; then smiting the surface, the thunderous concussion resounds for miles. You would almost think a great gun had been discharged; and if you noticed the light wreath of vapor from the spiracle at his other extremity, you would think that that was the smoke from the touchhole.

Fifth: As in the ordinary floating posture of the Leviathan the flukes lie considerably below the level of his back, they are then completely out of sight beneath the surface; but when he is about to plunge into the deeps, his entire flukes with at least thirty feet of his body are tossed erect in the air, and so remain vibrating a moment, till they downwards shoot out of view. Excepting the sublime *breach*—somewhere else to be described—this peaking of the whale's flukes is perhaps the grandest sight to be seen in all animated nature. Out of the bottomless profundities the gigantic tail seems spasmodically snatching at the highest heaven. So in dreams, have I seen majestic Satan thrusting forth his tormented colossal claw from the flame Baltic of Hell. But in gazing at such scenes, it is all in all what mood you are in; if in the Dantean, the devils will occur to you; if in that of Isaiah, the archangels. Standing at the mast-head of my ship during a sunrise that crimsoned sky and sea, I once saw a large herd of whales in the east, all heading towards the sun, and for a moment vibrating in concert with peaked flukes. As it seemed to me at the time, such a grand embodiment of adoration of the gods was never beheld, even in Persia, the home of the fire worshippers. As Ptolemy Philopater testified of the African elephant, I then testified of the whale, pronouncing him the most devout of all beings. For according to King Juba, the military elephants of antiquity often hailed the morning with their trunks uplifted in the profoundest silence.

The chance comparison in this chapter, between the whale and the elephant, so far as some aspects of the tail of the one and the trunk of the other are concerned, should not tend to place those two opposite organs on an equality, much less the creatures to which they respectively belong. For as the mightiest elephant is but a terrier to Leviathan, so, compared with Leviathan's tail, his trunk is but the stalk of a lily. The most direful blow from the elephant's trunk were as the playful tap of a fan, compared with the measureless crush and crash of the Sperm Whale's ponderous flukes, which in repeated instances have one after the other hurled entire boats with all their oars and crews into the air, very much as an Indian juggler tosses his balls.

The more I consider this mighty tail, the more do I deplore my inability to express it. At times there are gestures in it, which, though they would well grace the hand of man, remain wholly inexplicable. In an extensive herd, so remarkable, occasionally, are these mystic gestures, that I have heard hunters who have declared them akin to Free-Mason signs and symbols; that the whale, indeed, by these methods intelligently conversed with the world. Nor are there wanting other motions of the whale in his general body, full of strangeness, and unaccountable to his most experienced assailant. Dissect him how I may, then, I but go skin deep; I know him not, and never will. But if I know not even the tail of this whale, how understand his head? much more, how comprehend his face, when face he has none? Thou shalt see my back parts, my tail, he seems to say, but my face shall not be seen. But I cannot completely make out his back parts; and hint what he will about his face, I say again he has no face.

—Herman Melville,
Moby Dick

13
The tail of the whale: a pilot whale dives in Catalina Channel. (Photo courtesy of Randall R. Reeves)

How Do Whales Swim?

F.D. Omanney

Aquatic animals propel themselves through the water in various ways all of which, in fact, reduce themselves to the method of the paddle, the propeller or the jet. The jet method is used least but cuttlefish, squids and octopuses can dart backwards for short distances by squirting out a jet of water from the mantle cavity through a fleshy funnel. Diving birds such as auks and cormorants paddle with their feet and penguins with their powerful flipper-like wings. Fish and whales use the tail as a propeller. In fish it is triangular, crescentic or bladelike (eels) and vertically flattened. It beats from side to side, displacing water within the curve of the body out and back first on one side and then on the other so as to drive the body forwards. In whales the flat tail flukes are horizontal and beat up and down. They are built entirely of tendon and gristle, with no bones or any extensions of the skeleton such as we see in the tails of fish. They are well faired, as are the foreflippers, like the wings of an aircraft, a stout, smooth, rounded leading edge tapering finely to the trailing edge which is curved like a bow with a notch on the central axis of the body.

The upward thrust of the whale's tail flukes provides the forward drive, displacing water upwards and backwards. It has been suggested that the motion of the flukes has a rotary component but cine films of dolphins in tanks do not confirm this though they do show that the flukes have some independence of movement verticallly. The tail region of body moves up and down about a centre near the vent while the flukes make a similar movement about a second centre in the neck of the tail, so that the effect is that of a sculling oar.

Like the side fins of fishes, the flippers have a stabilizing and probably braking action but do not help to drive the body forward. The back fin, when present, is also a stabilizer. It is larger in whales that swim fast than in those that swim slowly, and indeed these may lack a back fin altogether. Unlike many bony fishes, but like sharks, whales cannot reverse and, with such comparatively small brakes, it must be difficult for them to stop.

Something that has long puzzled scientists in connections with the swimming powers of whales is known as "Gray's paradox." Dolphins can swim at speeds of 20 to 25 knots while the great whalebone whales can reach speeds of 15 knots. But the paradox is that experiments, conducted by Sir John Gray at Cambridge with rigid models, have shown that at these speeds the resistance of the water would be so great that the animal would need to develop as much as seven times the horsepower which it is known, from

---

How Do Whales Swim?
Reprinted from *Lost Leviathan,* by F.D. Ommaney, by permission of Dodd, Mead & Company.

measurements, to be capable of. In order to overcome the water resistance it would need muscles ten times more powerful than those of any other mammal. Yet this, it is believed, would be impossible because of the great heat which such high powered muscles would develop, and because the blood system could not supply the necessary oxygen.

The answer is thought to lie in the type of flow of water along the skin surface and in the reduction of what is known as "drag," the amount of resistance which the surface of a moving body offers to the flow of water past it.

Cine films have been taken of fish-shaped models towed through water rendered cloudy by a suspension of fine particles. They show that the particles in contact with the surface of the model form whorls and eddies which exercise a pull on the surface. This is called "turbulent flow." By using flexible models these eddies and whorls can be reduced and even got rid of altogether so that a condition known as "laminar flow" results. In this the particles slide smoothly over one another longitudinally without any pull against the surface of the model. It is believed that the skin of dolphins, and possibly of the great whales also, is adapted so as to produce this condition of laminar flow as the animal swims. American observers, watching dolphins swimming in tanks, have noticed that when the animal swims rapidly under water, certain folds and ripples appear on the surface of the skin at right-angles to the direction of motion, mostly on the lower and lateral parts of the body. They appear temporarily and last for

14
A sleek fin whale moves through North Atlantic waters. (Photo courtesy of Randall R. Reeves)

only a few seconds, and it has been suggested that they enable the skin surface to adjust itself to inequalities of pressure resulting from water resistance or drag. It may be, too, that the peculiar structure of the outer skin also plays a part in reducing drag. The skin of a dolphin, like that of all vertebrate animals, has a soft outer layer, the epidermis, and an hard inner one, the dermis. In most whales the epidermis is a very thin layer and the dermis is a much thicker layer of connective tissue underneath it. In the dolphins which the Americans had watched in the tanks, the Bottle-nosed dolphin, the outer epidermis is made up of innumerable fine ducts at right-angles to the skin surface filled with a spongy material. It is thought that this layer may be able to accommodate itself to minute oscillations of the water, microscopic eddies and whorls, caused by turbulent flow.

No such structure has been found in the skin of the great whales but on the lower surface of the body the skin is very flexible and the blubber is thin with loose connective tissue underneath it. It could therefore easily form folds and ripples, like those seen in the dolphin, when the whales swim rapidly under the water.

Another feature of the skin of whales is the fact that the two layers, dermis and epidermis, are closely dovetailed together by means of small papillae, minute protuberances, which are believed to counteract the sheering stresses set up by motion through the water. They are arranged along ridges of the dermis which somewhat resemble those we can see on our own skin, for instance in our finger prints and on the palms of our hands. These, too, serve to counteract stresses in grasping. On the whale's skin they are arranged in patterns, just as they are on our own, and it has been suggested that this pattern may correspond to the flow lines of water past the whale's body. The ridges tend to run obliquely form the lower side of the thorax up on to the back in the tail region. From this it has been argued that whalebone whales do not swim horizontally nose first but with head slightly raised and chest meeting the water first, as it were breasting the water, the flow past the body being, relatively, upwards and backwards. In this connection it has been noted that the whale's body is streamlined, or faired, on the back but not underneath except in the tail region where it is faired both above and below.

—F.D. Ommaney

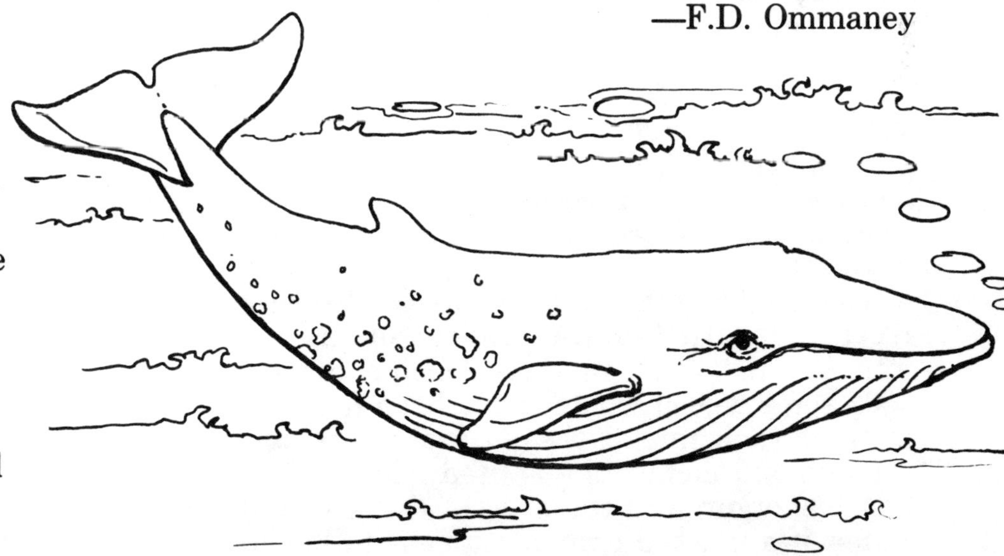

Exploring the Lives of Whales

Victor Scheffer

The author has spent 40 years studying whales and other sea mammals. He was with the U.S. Fish & Wildlife Service and was chairman of the U.S. Marine Mammal Commission, 1973-76. He is also the author of The Year of the Whale.

Just as zoologists have developed devices to eavesdrop on whale conversations, so are they always on the lookout for new methods of instrumenting whales — of fastening markers or recording devices to bodies that are smooth, flexible, and powerful. They have tried nylon girdles, which are prevented from slipping by a loop around the animal's back fin (an instrument is then fastened to the girdle). They have fed instruments, hidden in food, that radiobroadcast body temperature as they move through the gut. They are developing a tiny electronic device that can be surgically implanted under the skin of a captive whale slated for later release in the open sea.

The English cetologist Sidney G. Brown recently reported on the progress of an international whale-marking scheme. Since 1932 some 25,000 whales have been marked to track their migrations. At each marking a numbered metal tube is fired from a 12-gauge shotgun into the back of the whale. Later, if a marked animal happens to be killed, the mark is recovered at the whaling factory.

If whales are to be exploited—as they seem destined to be—this tracking of the great animals on their oceanic pathways is a key to their conservation. Whales must be managed, and a prerequisite will be to chart the boundaries of their separate populations.

In Seattle I watched a team of researchers fasten a radio transmitter to a 20-foot female killer whale, captured a month earlier in Puget Sound. The team secured the transmitter to the back fin with four bolts having zinc nuts that would corrode in seawater and eventually fall away, freeing the whale of her three-pound burden.

She and an 18-foot male companion, also radio-tagged, were released, and a tracking boat, the *Propeller*, followed their signals as they meandered generally northward some 75 miles a day. After nine days the *Propeller* lost them because of radio interference.

Few full-grown large whales have yet been radio-tagged. William E. Evans, of the Naval Undersea Center, once tagged a young gray. She was Gigi, captured in Scammon Lagoon, Baja California, tamed in a San Diego oceanarium, and released in the open Pacific in 1972, soon after her first birthday. For eight weeks she broadcast her position, as well as the depth and timing of her dives (maximum depth 558 feet; maximum downtime 16 minutes, 32 seconds). Then her radio fell silent, presumably because she broke off the antenna while swimming through kelp.

Most whales shy from human contact, but there have been exceptions, notably in San Ignacio Lagoon on the west coast of Baja California. In February 1976 a skiff carrying six whale-watching tourists entered the lagoon. As they drifted, a young whale, 28 to 35 feet long, slowly approached. Finally she bobbed up beside them.

Now I let Joseph G. and Dodie Vandeventer, from Sun City, Arizona, tell the story:

"She stayed near us, repeatedly rubbing her body gently under our skiff and surfacing. . . . For about 45 minutes this kept up, during which time all our party patted and rubbed her head and chin. . . . She was most curious and very friendly and interested. . . ."

Other friendly whales in San Ignacio Lagoon have captivated visitors: a few immature grays, including a couple playing at courtship, and even cows with calves.

—Victor Scheffer

15
How do you carry a whale? Ishmael the killer whale is lifted aloft on his way to a California lagoon. (Photo courtesy of U.S. Navy)

Seeds to Save a Whale

Maryann Dwyer

The search for a means to improve the economic conditions of American Indians living on poverty-stricken reservations in the Southwest may have important consequences for the survival of the endangered sperm whale. The seeds of a hardy desert shrub known as the jojoba (pronounced ho-HO-ba) contain an oil which can duplicate most of the industrial uses of sperm whale oil, the most important of which is as a high pressure lubricant. According to a report published by the National Academy of Sciences in 1974, during the last decade the United States consumed approximately 50 million pounds of sperm whale oil per year—a major factor in the whale's endangerment.

Flourishing in the wild in the extensive arid regions of Arizona, California, and Mexico, the jojoba reaches maturity in about five years, can produce about six pounds of oil for the 12 pounds of seed it yields annually, and has a lifespan of over 100 years. The oil of the jojoba seed is a unique substance—a colorless, odorless unsaturated wax which is easily removed from the seeds and readily usable in both its liquid and hydrogenated forms. In addition to its potential as a substitute for sperm whale oil, it also can replace beeswax and carnauba wax, whose prices have increased substantially in recent years. To top it off, scientists believe that the seed's shell also may be usable as cattle feed or "jojoba meal" if natural toxins in it can be removed successfully.

In 1969, enactment of the Endangered Species Conservation Act led to an end in U.S. taking of the endangered sperm whale and a cessation of sperm whale oil imports. Since then, dwindling industrial stockpiles and the absence of a completely satisfactory synthetic oil substitute have accentuated the need to develop an effective, inexpensive substitute such as the jojoba seed promises. Ironically, the existence of a federal stockpile of over 23,000 pounds being held in the event of a national emergency is adding to the need to come up with a substitute soon. The pressure is on for the government to sell its stockpiles to industry, and a bill in the House of Representatives, H.R. 3465, would permit that if it is passed. Some conservationists fear that once this oil is available, purchasers of it could mix it with illegally obtained sperm whale oil, which would not be easily detectable and thereby jeopardize sperm whale protection. However, if jojoba oil were available at competitive prices, industry would not be pressured to resort to illegal trade and further risk the survival of the sperm whale species.

Future jojoba cultivation as an agricultural crop will require a long-term investment of both time and money. Five years are required

---

Seeds to Save a Whale
By Maryann Dwyer in *Conservation News,* Oct. 15, 1975.

for a plant to mature before its seeds can be harvested. Since the jojoba now grows scattered throughout the regions of rough terrain in the Southwest, plantations must be set up so that the seeds can be mechanically harvested. Processing plants must be established to extract the oil from abundant quantities of seeds, convert it into marketable products, and utilize the wastes. Further experimentation is still necessary to determine the cross-breed which has optimal seed production as well as the best protection from predator and disease damage. Finally, identification of the full extent of industry's interest in jojoba as an oil substitute will be essential before any large scale production is initiated.

Presently, there are only a few experimental jojoba plantations in the world, two of which are funded by grants from the Department of Health, Education, and Welfare—one through the University of California at Riverside and the other by the Apache Marketing Cooperative Association in conjunction with the University of Arizona's Office of Arid Land Studies. The Association already markets its small harvest of jojoba oil in the form of candles, in addition to selling the oil itself and whole jojoba seeds. With the prices of natural and synthetic waxes on the rise, this may prove to be a small but lucrative business which is especially important to the government, since the first priority of HEW in awarding the grants is to aid Native Americans' economic development.

Jojoba oil is no miracle product, but it has great potential to remedy many current problems. Industry would have a viable alternative to sperm whale oil, synthetic and other waxes. There would be less pressure for the continued killing of the sperm whale. Native Americans would find in its production employment and economic security. And the U.S. would have a valuable new renewable natural resource.

—Maryann Dwyer, (1977)

A Catalog of Great Whales

Whales, large and small, along with dolphins and porpoises, make up the mammalian order Cetacea. Within the Cetaceans are two kinds of whales—the whalebone or baleen whales (Mysticeti) and the toothed whales (Odontoceti). Although sharing a common ancestor somewhere in the hazy past, the two suborders developed independently. There is more variety among toothed whales, who range from the small, playful freshwater and marine dolphins to the gigantic sperm whale. Baleen whales are more alike and more "whale-like"—and, today, far rarer.

BALEEN WHALES

Right Whales (Balaenidae). The right whales' name has been their undoing. The slow-swimming right whale, which floated after being harpooned, was the "right" whale to hunt for whale oil and whalebone. The right whales of the Bay of Biscay were wiped out several hundred years ago by Basque whalers; the Greenland right whale, or bowhead, was hunted nearly to death by 19th-century American and English whalers. Other whales of this family are the black right whale and the pigmy right whale. Right whales have long baleen plates, through which they strain plankton, and huge heads with arched upper jaws. Except for the pigmy, they have no dorsal fin. They can be spotted—by anyone lucky enough to see one of these rare creatures—by their V-shaped double spout. Right whales are all now protected from commercial fishing; the Eskimos still take a few.

Rorquals (Balaenopteridae). The blue whale, the largest animal on earth today—and perhaps the largest ever—can grow more than 100 feet long. The record weight measured is nearly 135 tons. Because of its size and strength, early whalers left the blue whale unmolested, but modern whaling technology has reduced its numbers from an estimated 100,000 in the early 1900s to probably less than 1,000. Some cetologists fear that the blue whale, even though protected, may never recover as a species. Blues usually travel in small, extended-family groups. They are monogamous, and a male Blue whale will stay with his mate if she is

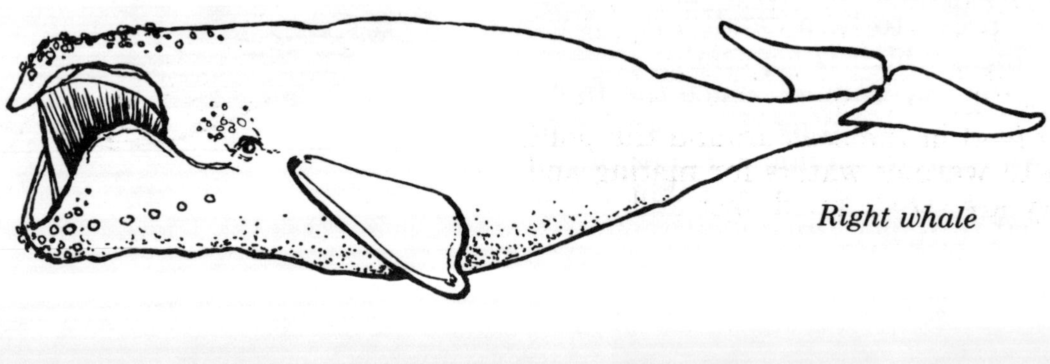

Right whale

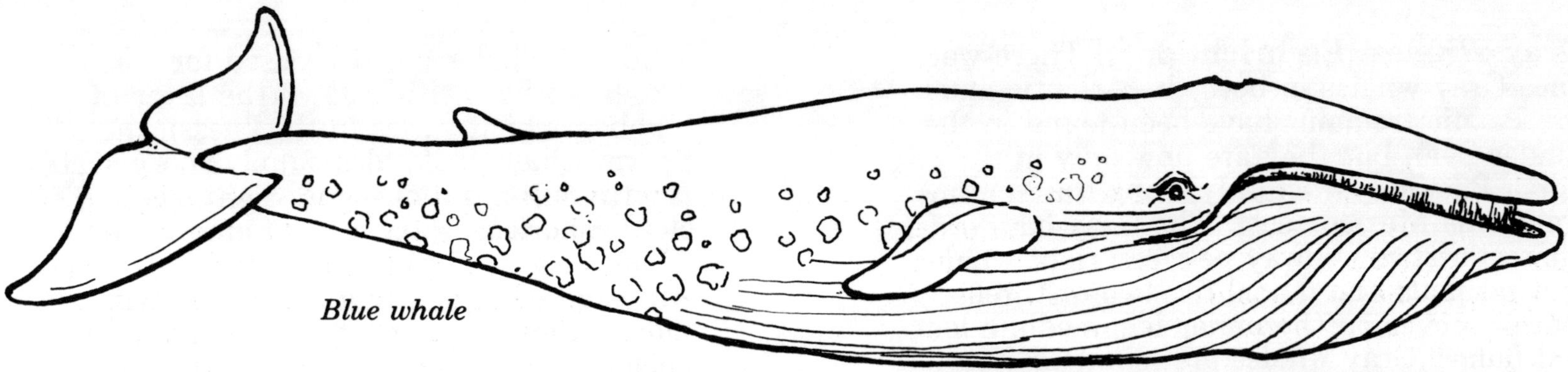

Blue whale

hurt. Females will stay to protect their calves. The Blues mate and bear their calves in warmer waters, but migrate to Arctic or Antarctic waters for summer feeding. Like other rorquals, they have short baleen plates, which serve as strainers to filter krill (tiny, shrimp-like animals) from the water. Whalers sometimes called blue whales "sulphur-bottoms," because of a yellowish cast given their undersides by tiny diatoms growing there.

Second in size to the blue whale among rorquals is the Finner, or Fin whale. It averages about 60-65 feet long. Finbacks are asymmetrically marked with white on the underside. Next in size are the Sei and Bryde's whales (about 60 and 50 feet respectively). The smallest rorqual, the Minke, is only about 30 feet long; it is also called the piked or lesser rorqual. All the rorquals have deep "pleats" or grooves on their undersides, giving them their name ("tubed whale" in Norwegian). Deep-water whales, they feed in the seas around the poles and migrate to warmer waters for mating and calving. They are streamlined, fast and graceful.

An oddity among the rorquals is the Humpback whale, which is sometimes classified in a family by itself. It is chunky rather than streamlined, with fewer "pleats" on its underside. Humpbacks average about 40 feet in length. They, too, migrate from summer feeding grounds near the poles to warmer waters. Humpbacks have a reputation for playing boisterously, leaping and breaching with clear enjoyment. Most distinctive of all, the Humpbacks—as far as we know—are the only whales that sing.

Humpback

Gray Whales (Eschrichtidae). There were once Gray whales in both the Atlantic and the Pacific (remains have been found in the Zuider Zee), but they are now only in the north Pacific, particularly along the California-Mexico coast. The Gray has characteristics midway between right whales and rorquals—no dorsal fin, two or three short grooves on the underside, medium-long tail flukes. Gray whales are about 45 feet long, weighing about 20 tons. Intensely hunted in the late 1800s, the California Grays drastically declined in population, but are slowly recovering. There are probably 10-12,000 now.

TOOTHED WHALES

Sperm Whale (Physeteridae). Among toothed whales, only the Sperm whale ranks as a "great whale." As the carrier of three economically valuable products, it was the prime target of the New England whalers.

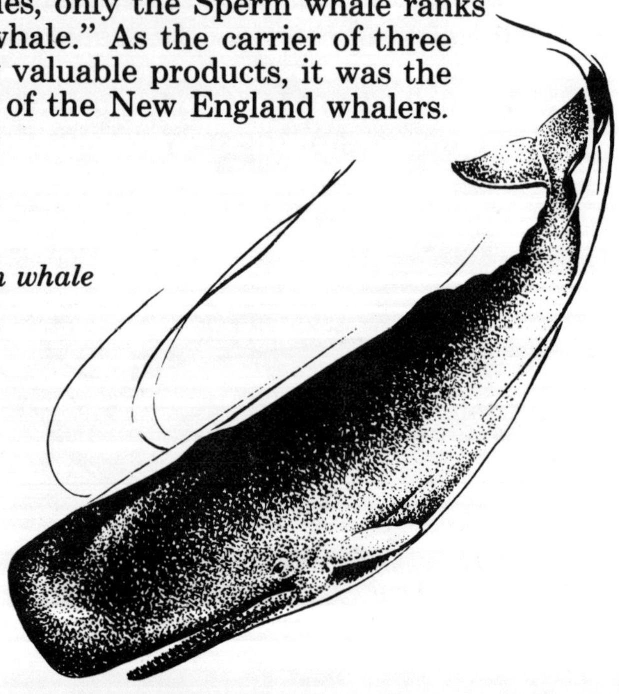

Sperm whale

While all whales could be used for whale oil (produced by melting down the layer of blubber, which varies in thickness), the Sperm whale could also supply spermaceti (sperm oil) and ambergris. Spermaceti, a clear, colorless, waxy oil, is found in the whale's gigantic, squared-off head. Ambergris—evidently a product of whale indigestion—was immensely valuable for perfumes.

Unlike the essentially monogamous, small-family baleen whales, the sperm whales form harem groups around a dominant male. Bachelor males also travel together, as do loose groups of lone bulls (rather like the organization of an elephant herd). Moby Dick, a white sperm whale, was probably a rogue male. Sperm whales feed mainly on squid and cuttlefish; stories of battles between gigantic sperm whales and giant squid are not just tall tales, for many sperm whales carry the scars of huge squid tentacles.

To the Greeks and Romans the sperm whale was *physeter*, and the French term *cachalot* was also widely used. The most abundant of the great whales, sperm whales have also declined because of hunting. One population estimate is about 250,000. There are also two species of pigmy sperm whale, called Kogia—whose 12-foot length makes them pigmies indeed beside a 33-ton, 60-foot-long male sperm whale.

Bottle-nosed and beaked whales (Ziphiidae). The largest member of this family, the giant bottle-nosed whale, reaches about 40 feet in length. Like its larger cousin, it has a reservoir of waxy spermaceti in its head. Although these are toothed whales, many have no visible teeth in their jaws. Some, like the Tasmanian beaked whale, are very rare. The family also includes Cuvier's whale, the cowfish, and Sowerby's whale.

16
Bottlenose dolphin in the North Atlantic. (Photo courtesy of Randall R. Reeves)

White Whales; Dolphins and Porpoises (Delphinidae). The beluga and the one-horned narwhal make up one group of this family. Killer whales (orca), false killer whales, and pilot whales (blackfish or grampus), along with dolphins and porpoises, are the other subgroup.

Lake and River Dolphins (Platanistidae). This group includes many smaller, mostly freshwater species of dolphins.

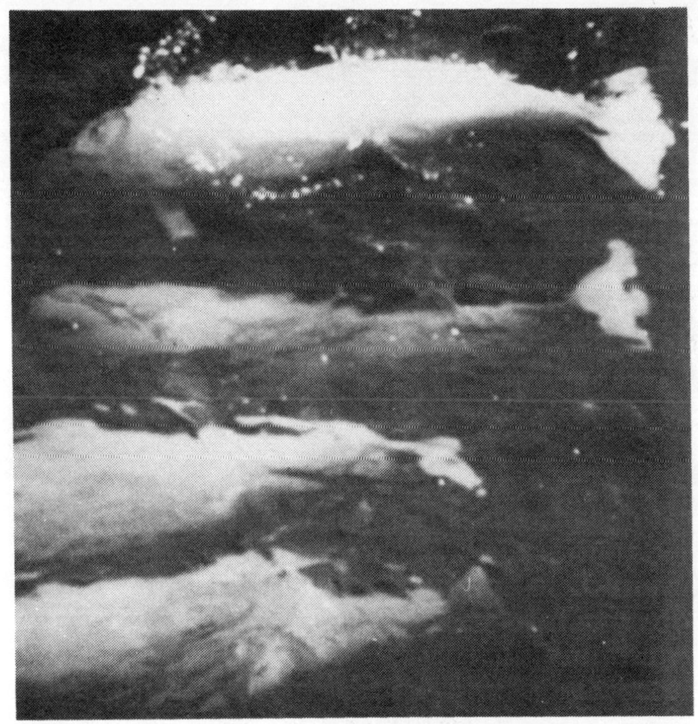

17
A school of belugas traveling the Saguenay River. Three are beneath the surface. (Photo courtesy of Jean Laurin)

18 A fin whale in the North Atlantic. (Photo courtesy of Randall R. Reeves)

3

THE ENDANGERED WHALE

The only creatures on earth that have bigger—and maybe better—brains than humans are the Cetacea. . . Perhaps they could one day tell us something important, but it is unlikely that we will hear it, because we are coldly, efficiently, and economically killing them off.

—Jacques-Yves Cousteau

Whale Watchers: Job

Canst thou draw out Leviathan with an hook? Or his tongue with a cord which thou lettest down? Canst thou put an hook into his nose? Or bore his jaw through with a thorn?

Will he make many supplications unto thee? Will he speak soft words unto thee? Will he make a covenant with thee? Wilt thou take him for a servant forever?

Wilt thou play with him as with a bird? Or wilt thou bind him for thy maidens?

Shall the companions make a banquet of him?

Shall they part him among the merchants?

Canst thou fill his skin with barbed irons?

Or his head with fish spears?

Lay thine hand upon him, remember the battle, do no more.

(Photo courtesy of Randall R. Reeves)

The Star-Crossed Whale

Wesley Marx

A whale is a living monument to the grandeur of the ocean. A creature of the land by heritage, the whale inhabits the ocean by choice. On land a whale would be imprisoned by gravity, an easy target for predators one-tenth his size. Only the ocean can provide it with the support to buoy its massive size, the food to support its appetite, and the space to accommodate its movements. On land the whale would be as quick as a snail. In the ocean the world's largest mammal can pace an ocean steamer.

Without the whale the ocean would be deprived of its most lordly symbol—and deprived it may well be. The threat to the whale from commercial killing is relatively well known. Yet like the American Indian of the nineteenth century, the whale can face extinction not only by loss of life but by loss by living space. These twin forces of extinction focus with special intensity on a whale species of unique value to man.

Baja California dangles boldly from the southwestern extreme of the continental United States, a geological dagger that plunges into the Pacific and splits the ocean apart from the Gulf of California and northern Mexico. The Sebastian Vizcaino Desert covers the middle of this dagger (160 miles longer than the Italian boot) like rust. Through niches chiseled by combers, the Pacific spills into this desert to form a series of ocean lagoons.

Scientists are not sure of all the reasons that attract the gray, a large whale species forty tons in weight and up to fifty feet in length, to Scammon Lagoon, but they feel that they understand the main ones. To begin with, the lagoon is of a size to accommodate the two thousand or so whales that come annually; a two-mile wide ocean entrance expands into a body of water some thirty-five miles long, up to eleven miles wide, and one hundred twenty feet deep. Instead of a lagoon, this vast body of water resembles a biblical sea that would merge with the sky except for a thin desert horizon. Yet the lagoon, despite its size, generally remains as placid as its counterpart in a municipal park. This can be important to a pregnant whale, who must teach her one-ton baby to breathe and swim immediately. Although these lessons can be learned on the open ocean, swells can complicate matters. The lagoon's high salinity—twice that of the open ocean—provides another benefit, added buoyancy.

Dr. Theodore Walker of Scripps Institution of Oceanography believes that eel-grass beds are another lagoon attraction. Dr. Walker believes that the jaws of the whale scoop up these beds, which are then flushed through the whale's baleen—brushlike filter gums—to

---

The Star-Crossed Whale
Reprinted by permission of Coward, McCann & Geoghegan, Inc. from *The Frail Ocean* by Wesley Marx.

Plate V

C.M.Scammon.del. Lith.Britton & Rey S.F.

CALIFORNIA GRAYS AMONG THE ICE

19
(Engraving from Scammon's **Marine Mammals of the North-Western Coast)**

expel mud and water and retain shellfish and other eel-grass morsels.

In addition to breeding and feeding attractions, the lagoon offers the gray whale seclusion. Man-made intrusion has caused it to desert its ancient breeding grounds. In Scammon the grays have enjoyed a natural solitude protected by the lagoon's remoteness and inaccessibility to man—at least up to the present.

The whales that enter this desert-flanked lagoon come all the way from their choice feeding grounds in the Arctic Ocean, a migration of some six thousand miles, or twelve thousand round trip. (A smaller herd summers in the Sea of Okhotsk and winters in the bays of southern Korea.) Despite their massive bulk, restricted vision (whales cannot even see their tails), and impotent sense of smell, the gray whales succeed in making this magnificent migration year after year.

The migrating gray's ability to overcome such natural deficiencies testifies eloquently to his adaptive nature. To make up for deficiencies in sight and smell, the gray utilizes a batlike sense of hearing and organic sonar (echo sounding) that enables him to "see" through sound and navigate the northern Pacific Ocean.

Whale hydrodynamics consists of an economy of motion. The huge bulk is streamlined into an elongated exclamation point with such friction-free features as recessed nipples, ears, and reproductive organs. The streamlined whale has his nostrils on the top of his head. As a result, he never has to thrust his ponderous head out of the water to breathe. Explains Dr. Walker, "The whale comes up from the depths to breathe like a jet coming down to land, everything timed so that only the blowhole breaks through the surface." The whale's flukes, or fanshaped paddle, drive this streamlined body at speeds as high as fifteen knots, a good pace for any large ship.

Equipped with a keen sense of hearing, superb hydrodynamics, and organic food storage, the Arctic-grazing grays head for their desert birthplace in mid-September, followed for a short time by white snowbirds which gorge on their leavings. The grays are a loose, straggling lot in migration, separated by miles of ocean and weeks of time. In the lead are the pregnant whales, which set a pace of about one hundred miles a day. This informal procession passes through the Chukchi and Bering seas before the cold waters stiffen to ice, loops around the Aleutian Islands, and trails along the coast of British Columbia. Occasionally some members of the procession pause to rub their barnacle-encrusted torsos against the anchor chains of a Canadian lightship.

By November the procession passes within sight of Cape Flattery, a lonely headland in Washington. Countless blows, or "plumes," form rainbow prisms in the light of the ocean sun. Soon yachtmen out of San Francisco can watch certain whales exchange taps with their front stabilizer flippers. Because these exchanges occur between males and eligible females, they are regarded as love taps. They resound for miles.

Occasionally, a group of grays will hastily

meld into a giant, revolving pinwheel with their limber tails as spokes. This formation wards off the wolflike packs of toothed killer whales. (The U.S. Navy once depth-bombed killer whales who were swallowing Iceland's fishing economy.) That is another of Scammon Lagoon's attractions: an absence of the swift killer whales.

By late December this rainbow-plumed procession tightens up and passes in review off Point Loma, a San Diego headland. About three hundred miles south of Point Loma and three weeks later the lead pregnant whales come within echo sounding of submerged sandbars. "No definite directions can be given for crossing the bar at Scammon, which is constantly changing," notes the U.S. Navy Hydrographic Office. Like phantom beaches, these submerged bars trip up the offshore swells and create a surf in the middle of the rolling blue sea. They guard the entrance to the placid lagoons of Baja California and tell the gray whales that they have reached their destination. The grays, with a draft of six to eight feet, thread their way through the submerged obstacle course.

The grays have three major breeding areas: San Ignacio Lagoon, the lagoon channels in Magdalena Bay, and Scammon Lagoon. Scammon attracts the largest number, approximately two thousand. During January and February they are so profuse that their constant blows endow the lagoon with a built-in sprinkler system. If you venture into this whale lagoon in a boat, you will find that the whales become navigational hazards. In the near distance a glistening rubbery hull rends the calm waters, spouts a minor geyser of condensed breath, and slips silently back beneath the blue cover of the lagoon. Without warning, but fifteen yards away, a sonorous exhalation sounds, like the whoosh of a steam press, and a blowhole, spewing like a volcanic island, emerges. A cloud of vapor vanishes, exposing the two dark pits that mark the nostrils, and a long grayish-black hull crusted with knobby barnacles trails the submerging blowhole. You watch, rooted in amazement, as the mere backbone of a whale instantly cuts everything in the world down to size. Then, to the rear, a smaller blowhole emerges, spouts, and sounds. A newborn calf is following the lead of its mother.

Sometimes a whale's mighty forward torso will shoot out of the water like a metal pile and then topple in an explosion of foam. This maneuver is called "breaching." Another whale will emerge from the lagoon depths in what is called a "spy hop" as casually as a cobra from a basket. His lower jaw unhinges like a drawbridge and the desert sky becomes pinioned in his open mouth. The jaws close slowly on the sky and the head sinks back beneath the water.

The terms that designate these impressive maneuvers are merely descriptive. Scientists differ on what, if anything, is being accomplished. Dr. Walker connects spy hopping with feeding. "It is my theory," says the Scripps scientist, "that the whale assumes this vertical position—sometimes with its flukes (tails) actually on the bottom to help support it—because this posture brings the heavier seafood to the bottom of its throat where it is easier to swallow. The throat muscles can actually be seen swallowing." Some scientists speculate that spy hops are a form of male chauvinism, for they occur repeatedly in one narrow neck of the lagoon. This particular body of water, referred to by scientists as Fornication Hole, is where whales mate. Here whale taps thunder and whale heads collide affectionately. During this foreplay the male's eight-foot penis emerges from its recessed pocket. The courting pair then merge close to the surface and begin revolving like a cumbersome lathe. Whales are polygamous and a high level of activity is maintained in the mating area.

When an impregnated female returns to Scammon the following year, she passes through Fornication Hole with a one-ton baby in her womb. She cruises into more spacious waters called the Nursery. Gulls, which congregate to feast on the afterbirth, indicate to human observers when the underwater birth has occurred. In a frenzied cloud of white feathers, a new mother will lift her head clear of the water. On top is the baby, some fifteen feet long. The mother's head submerges and re-emerges with the baby whale perched in the same place. The mother is teaching her newborn to breathe. Otherwise, two round-trip migrations, twenty-four thousand miles in all, will go for naught. After a while the calf will nestle close to its mother's underside near the surface. The recessed nipples form an airtight connection with the calf's mouth which shuts out the salt water. To expedite the feeding the calf does not suck. The mother squirts a gallon or two of milk down the calf's throat.

After six weeks in the Nursery the mother leaves Scammon, trailed by her calf who will be fully weaned in the microshrimp pastures of the Arctic Ocean. It will be the first of many migrations for this baby whale. How many is not known because scientists do not know how long whales live.

The gray's affinity for coastal waters, in particular a desert-flanked lagoon, is unique in whaledom. Other large whale species steer clear of coastal waters; on leaving the freezing polar seas in the fall, they seek out island sanctuaries—the Azores, Malagasy, Melanesia, New Zealand—to winter in uninhabited inlets and atoll lagoons. Precisely because of its affinity for coastal waters, the gray becomes a bellwether of the whale's ability—or lack of ability—to survive the forces of man.

In the nineteenth century Yankee whalers ventured into the open seas to track down the big blubber whales, following creatures faster and larger than their own ships through icy gales, dead calms, and uncharted waters. In January of 1846, a New England whaling ship en route to the polar grounds of the prized bowhead whale came to anchor in Magdalena Bay. The astonished crew saw countless whale spouts pricking through the calm blue surface of the bay lagoons. A whale-killing spree promptly ensued. What was patronizingly dubbed the "mudhole season" came into being. This patronization obscured a salient trait of lagoon whaling. On the open seas, a whale—once he was harpooned—tended to run and sound in the hope of shaking loose the whale boat that clung fast to the end of the harpoon line. The shallow, confined nature of the lagoon thwarted this running tactic and forced the gray whale to meet his tormentors head on. Because the grays were more agile than the whaling boats, the pursuers suddenly found themselves the object of pursuit. Sometimes the harpooned gray would corner his tormentors and ram their wooden craft. Sometimes his mighty flukes would clear the water and crash down on the luckless whaling crew in a tactic called "lobtailing," and sometimes the line on a boat made "fast" to a whale would suddenly go slack. The perplexed crew would look over the side of the boat to sight the whale, but the waters would be murky with mud and sand churned up by their quarry. Then, in a frightful flash, the crew would be lifted clear of the lagoon and sent flying through the air. The whale, playing possum, simply waited for the lagoon current to bring the whale boat overhead . . . and into position for a reverse lobtail. "I shipped to go a-whaling," complained one surviving lobtail victim to his captain. "I'd no idea of bein' required to go into a duck pond to whale after spotted hyenas." The gray's deadly cunning resulted in a bad press and temporarily discouraged exploitation of the whale ponds, but commercial pressure on whalers to confront the "devil fish" increased.

In 1857 the brig *Boston* entered a previously ignored lagoon. On the debut of whaling in its waters, a jubilant captain, Charles Melville Scammon, could report, "Two large cows were captured without difficulty, which gave all hands confidence in our ultimate success." This confidence crumbled the next day. Before a whale could even be struck (harpooned), two of the *Boston's* three whaling boats were staved in by a succession of flashing flukes. Scammon found himself in command of "a crowded and contracted hospital. Our situation was both singular and trying. The vessel lay in perfect security in smooth water; and the objects of pursuit, which had been so anxiously sought, were now in countless numbers about us."

One morning a boat was lowered from the "contracted and crowded hospital." The crew carefully hugged the shallow lagoon shoreline, out of reach of the whales, and finally anchored inshore where the lagoon narrowed into Fornication Hole. When a whale emerged to spout, a man in the boat shouldered a new form of technology. The bomb-lance gun fired a bomb lance designed to pierce the whale's blubber and timed to detonate inside the whale's lungs. It had one drawback—a short range. When the the whale emerged again, the man aimed and fired. When the color of the whale's spout turned crimson, the man knew he had successfully pressed the gun's range.

On the open seas such marksmanship would have been futile. A "bombed" whale would sink or, once it floated to the surface, be out of sight. Scammon hoped that the lagoon, by its largely enclosed nature, would make the bombed whale fast. The following day, the whale was floating near the head of the lagoon, buoyed up by gases decomposing in its blubbery hull. Scammon was no longer in the business of pursuing whales. He was

20 *Whalers with a "Greener's Gun" mounted on their whaleboat. (Drawing from Scammon's* **Marine Mammals of the North-Western Coast)**

ambushing them.

In one season Scammon had not only discovered a prime breeding ground of the gray whale but had devised a means of making the defiant grays as defenseless as steers in a stockyard. Bustling boat crews knitted shut the giant lips of dead whales and towed the bloated corpses to the trying pots. Stinking smoke from the trying pots spiraled into the desert sky accompanied by the babble of Portuguese, Kanaka, and Yankee tongues.

A sudden influx of whalers fostered further refinements on the lagoon slaughter. A whale boat would chase a stray calf into the lagoon shallows. Soon the anxious mother would appear to retrieve her young, only to strand in the shallows. Once the mother was caught fast by the low tide and her own frantic exhaustion, the whaling boat would reappear for the kill.

Lagoon whaling was soon practiced just as intensively in other lagoons, small and large, along the Baja coastline. While this maritime carnival reveled in the gray's breeding grounds, eleven whaling stations along the California coastline bushwhacked grays en route to the lagoons. This commercial aggression could have but one end. By 1861, four years after Scammon sailed into his lagoon, gray whaling was no longer economical. Too few grays were left. Since the initiation of lagoon whaling an estimated ten thousand eight hundred whales had gone into the trying pots. This mortality did not include the calves who endlessly circled the ships where they had last seen their mothers until starvation set in. The whalers departed from the desert lagoons to find new grounds and to apply new techniques. From a prewhaling population of some thirty thousand, the grays descended to "no more than a few dozen," according to one San Diego naturalist who missed their lordly procession off Point Loma in the 1930's. Ironically, the grays' vulnerability to whaling suddenly contributed to their survival.

In 1938 the International Convention for the Regulation of Whaling instituted a voluntary method of conservation quotas on whale kills. The convention also accorded the gray a special status in recognition of its plundered condition. "It is forbidden to take or kill gray whales except when the meat and products of such whales are to be used exclusively for local consumption by the aborigines."

With Scammon Lagoon off limits to whaling captains, the grays have brought their population level up to an estimated six thousand, but the beneficent clause of the International Convention protects them only from commercial exploitation. It does not protect their vital living space, and thus they once again face the threat of ultimate extinction.

The Vizcaino Desert helps to make Baja California one of Mexico's least populated regions, but it does manage to host one small community, a prosperous one at that. Guerrero Negro is surrounded by the source of its prosperity, giant solar salt ponds that stretch out of sight like snow-covered prairies. The saline-rich waters of Scammon Lagoon which buoy up baby whales are pumped into these ponds at the rate of 387 million gallons daily. To up annual salt production from one million to three million tons, Exportadora de Sal, owner of the 300,000-acre salt concession, is in the process of dredging part of the lagoon as a loading dock and a salt barge channel. Barges will transship the salt harvest to 100,000-ton tankers which can not worm their way through submerged sand bars.

Exportadora, which spent sixteen million dollars to construct the saltworks complex, will spend an additional twenty million to realize expansion plans. The prime obstacle to such ambitious plans ordinarily would be financing, but not in Exportadora's case. Exportadora is an affiliate of National Bulk Carriers, Inc., of the United States, which in turn is controlled by D.K. Ludwig of New York. Besides owning outright the world's pioneer fleet of supertankers, Mr. Ludwig dredges the Orinoco River in Venezuela to accommodate his iron-ore tankers. (Recently he wanted to dredge a twelve-mile channel through the Everglades country of Dade County, Florida. The dredged-up portion of Everglades, some forty-five million cubic yards in all, would provide earth fill for the site of a petrochemical-harbor complex, but Mr. Ludwig could not obtain a zoning variance for such nature moving. Wildlife groups and Miami tourist interest, rallied by resident-author Philip Wylie, successfully protested that the price of the channel project would be the scenic beauty and abundant life of the Everglades.)

Although the fourteen hundred residents of Guerrero Negro, whose homes, shops, and gypsum roads are owned by Exportadora, are understandably happy over the expansion plans, scientists who faithfully trek to Scammon Lagoon each January to document the gray's return from the "extinct and vanishing category" are much less so. In 1960, before Exportadora began turning towards Scammon, Dr. Raymond Gilmore, conductor of a gray-whale census for the U.S. Fish and Wildlife Service, wrote, "The area available for calving and mating may be the most critical factor in determining eventual size of population. Reduction of this area by natural events or by man may have a pronounced effect in lowering the population ceiling." Today Dr. Gilmore sees in the plans of Exportadora the seeds of just such a "pronounced effect."

Such scientific premonition is based on the nature of the whale and a reading of history. The whale's craving for seclusion, as offered by Scammon Lagoon, is no mere whim. In "seeing" through sound, whales can be temporarily blinded by distracting noises, such as the hum of a ship's screws, the moan of harbor dredges, or the clatter of loading docks. The century-old desertion by the grays of San Diego Bay as a winter home coincided with harbor development. A hundred years later, Dr. Gilmore sees the whale's low tolerance to noise re-enacted in Magdalena Bay, where increasing fishing-boat traffic caused the "gray whales to consort in numbers only in the inaccessible tributaries." (Magdalena Bay is now being transformed into a five million dollar deep-water port and seaside playground.)

Oily barge emissions raise another discouraging possibility: eventual pollution of the eel-grass beds.

Although initial barging activity at Scammon will occur in a lagoon arm not regularly frequented by the whales—a company decision based on engineering considerations, not conservation—there is no specified end to lagoon "development" (The narrow channel through the shifting sand bars will foster whale-barge encounters outside the lagoon.) With whale requirements at the mercy of salt requirements, an eviction of whales from Scammon Lagoon becomes possible, either gradually or quite suddenly. Faced with a continuing takeover of its winter grounds in Baja California, the gray will have to find new coastal breeding and feeding grounds, compete with other whale species for offshore grounds, or subsist on the open ocean. The uncertainty of such adaption intensifies scientific concern.

Such a prospect has a particularly galling significance to scientists. Just like commercial whalers, highbrow whalers must generally take to the open seas to track down the object of their professional curiosity. But observing a whale from a ship—if one is sighted—is like observing an iceberg; most of what matters is not visible. As a result, even though factory ships know how to slaughter whales on an assembly-line basis, scientists know distressingly little about a creature that pumps blood through hundreds of feet of arteries at one pulse, maintains a body temperature of ninety-six degrees in near freezing waters, manufactures a mother's milk twice as rich as that of human mothers, and possesses a sponge-like bone structure that can redeem multifractures.

The almost unbelievable opportunity that Scammon Lagoon provides—a pondlike atmosphere in which to observe the world's largest mammal intimately—was first recognized by Captain Scammon, incredibly enough. While ordering harpooners to bomb whales, the self-educated Scammon was also measuring the girth of the victims, inspecting the contents of their stomachs, and executing precise drawings of their magnificent torsos. Like another whaling contemporary, Herman Melville, Scammon sensed that there was more to whales than oil to light cities and bone to corset can-can lines. In 1874 he made what the magazine *Science* soon declared was "the most important contribution to the life history of marine mammals ever published": *The Marine Mammals of the North-western Coast of North America.*

With the gray whale in the almost extinct category around the turn of the century, Scammon Lagoon slipped from scientific notice. In 1926 a famous Hollywood actor was cruising through the lagoons of Baja California in his yacht *The Gypsy.* John Barrymore was entranced by two gray whales in "amorous dalliance . . . if anything so enviably and titanically active comes under the heading of dalliance." He whipped out his camera. "We eavesdropped, and I trust, for the stimulus of one's old age, recorded the largest mammals in flagrante delicto! A memorable day!" Barrymore, who harpooned a gassy rubber facsimile of a whale in the movie "The Sea Beast," then tried to accomplish the real thing on a gray, but failed, according to his biographer Gene Fowler.

Errol Flynn, a good friend of Barrymore, was thrilled by these encounters. Shortly after World War II the swashbuckling actor, son of an Australian biologist, personally chauffeured a Scripps scientist to Scammon in his somewhat infamous yacht. Dr. Carl Hubbs, an ardent reader of Captain Scammon's lifework, confirmed that the gray whales existed in goodly numbers in the lagoon. This confirmation has since sparked eager scientific pilgrimages.

These pilgrimages have often been confined to testing, and sometimes discarding, techniques designed to overcome the whale's size and surprising sensitivity. Dr. Paul Dudley White, heart specialist to Presidents, thought it would be instructive to record the heartbeat of a whale. He set off for Scammon Lagoon with an electronic harpoon, a harpoon crew, and a swift outboard. Unfortunately, Dr. White's crew drifted between a mother whale and her calf, and the mother nearly staved in their boat with a flick of her flukes. Dr. White then resorted to a hovering helicopter, but the grays proved too agile for aerial harpooning.

Scammon Lagoon also attracts acoustical physicists, armed with buoy hydrophones, who seek to record and identify whale sounds. It seems that the U.S. Navy cannot always tell a whale sound from a submarine sound, and this is not a healthy situation, either for the Navy or the whale. Lagoon whales, whether or not they know they are being bugged, tend to remain silent in the vicinity of buoy hydrophones. Two physicists were reduced to writing a scientific paper entitled "The Silent Whale." "Submarines go silent when they are being spied on," suggests Dr. Walker, "why not whales?"

In his study of whale behavior Dr. Walker generally scorns exaggerated technology. On occasion he has used a rowboat topped with a plywood forecastle that looks as if it were recruited from the rental stock of a park pond. With the motion of the lagoon tidal current, Dr. Walker likes to creep up on a forty-five-foot whale until its occasional blow virtually clouds in *Walker's Ark*. "Creeping a whale" gives Dr. Walker intimate glimpses of bull whales napping and cow whales feeding, but unnerves guest observers, including myself, who have seen a whale pass beneath the *Ark* like a silent freight train.

As oceanographers like Dr. Walker comprehend the organic system that sustains the world's largest mammal, other oceanographers speculate on how this knowledge might be utilized. In "Ocean Engineering," a voluminous study by North American Aviation, the high cost of obtaining and equipping vessels specific to collect oceanographic data is discussed. The placing of sensors for temperature, salinity, and other factors on wide-ranging marine animals is posed as a possible solution. The study notes, "As an example, the California gray whale ranges from Lower California, up the Pacific Coast to Alaskan waters. . . . At the same

21
Two whales in the Catalina Channel. (Photo courtesy of Randall R. Reeves)

time, it moves from the surface to great depths; thus, an animal is available to randomly probe a large volume of the Pacific waters. It also surfaces several times an hour and could, therefore, telemeter recorded information at that time. If many ships were provided with simple recording and automatic direction finding equipment, the data transmitted could be reduced to a practical form on an almost continuous basis." This proposed solution ends with a note of caution. "The hazards of working with this animal are not to be underestimated, however."

As a scientific whale observatory Scammon Lagoon could in time become the place in which the whales could reveal their natural secrets to man. As a general whale observatory it could become the place in which the gray whale could display its grandeur to the world. Already the gray is a major tourist attraction in southern California, an area in which the competition is fairly hectic. Each December the National Park Service operates the world's pioneer whale observatory on strategic Point Loma. In addition, three San Diego sport-fishing firms operate "whale-hunt" boats, which put passengers "practically out on the backs of whales," according to Dr. Walker. These passengers pay just to see the backs and blows of whales. Yet three hundred miles to the south they could watch whales breaching, spy hopping, sleeping, nursing, and mating, just from the shoreline.

As roads and modest tourist accommodations penetrate Baja California's scenic remoteness Scammon Lagoon could achieve its potential as a world-famous whale observatory—*if* rigorously supervised. Water ski rentals, whale-hunt boats, and sewage outfalls could chase away the grays even more quickly than plodding salt barges. Governed properly, Scammon Lagoon might well function as both a salt-barge channel and an observatory. Although scientists believe that whales have a low tolerance for man-made activity, they do not yet know the threshold levels of this tolerance. If such levels could be established by a program of yearly monitoring at Scammon, regulations on barging activity might be formulated. That the whales occupy Scammon but two months of the year could facilitate acceptance of these regulations.

Mexico possesses the authority to implement such a program. Yet conservation in a developed nation is difficult enough to achieve; in a developing nation it is an even more trying task. Just as it was in San Diego Harbor a century ago, the plight of the gray whales can be totally obscured by immediate considerations of economic development, for when the government of Mexico thinks of Scammon Lagoon it must consider Exportadora de Sal as well as whales. For a country striving to improve the lot of its people, Exportadora has colonized a desert area, installed an ocean harbor, created an important export commodity, made Mexico the third leading salt producer among developing nations, and now promises to equal that achievement among the developed nations as well. Unlike Dade County, Florida, Mexico is not particularly inclined to interfere with the visions of Mr. Ludwig.

Although signatory to the International Convention that protects the grays, Mexico has considered opening up whaling on them. In 1965 Mexico asked for and received bids for a whaling station that would convert lagoon whales into oil and whale steaks. Although inopportune from the viewpoint of the convention or conservation, such a station might provide a quick financial killing. The harpoon cannon and the factory ship have made the mighty blue whale as vulnerable to purges as the gray, so vulnerable that whaling is at its lowest ebb. Two firms from Japan, a large whaling nation (also signatory to the whaling convention), are among the eager bidders, even though the gray-whale stocks cannot ensure a long commercial life for the station. The plans have not yet come to fruition. If they do, the whalers of the United States, Canada, and Russia cannot be expected to sit idly by as the grays migrate down to a Mexican whaling station.

One agency in the Mexican government now appears to dissent from the proposal. In the fall of 1966 Dr. Enrique Beltran, director of the Mexican Institute of Renewable Natural Resources, forwarded to me a letter he received from Jorge Echaniz, Mexico's General Director of Fishing. Mr. Echaniz said his department was drawing up a "recommendation for the protection of this species (the gray whale) which will define our opinions on this matter, essentially pointing up the necessity of building a sanctuary in Laguna Ojo de Liebre (Scammon Lagoon)." Mr. Echaniz expressed concern that

Exportadora boats had already caused whales to move away from Guerrero Negro Lagoon. He also indicated concern over gray whaling. "No path should be left open for its possible exploitation, since it has taken thirty years for the species' recuperation, and it could be easily lost by a brief lapse into irrational capture."

If such views are politely filed away, Mexico's economic ambition will join all the other forces that bear down on the gray whales of Scammon Lagoon with the intensity of a conspiracy—the world-wide industrial appetite for salt, the solar-salt qualities of the Vizcaino Desert, the business acumen of D.K. Ludwig, the evolution of supertankers, the muscular technology of deep-water dredging, and the weakness of whaling conservation. The main ally of the whales, inaccessibility, crumbles before these forces. Only man himself can provide the protection that this ally once offered. Dr. Gilmore, who has watched the grays recover from one disaster, concludes, "The gray whale is a relict species in a shrinking environment."

Because other large whales steer clear of coastal waters is no guarantee that man's industrious nature will bypass their grounds. Smaller whale species who do favor coastal haunts have already felt the brunt of this shrinking environment. For three hundred fifty dollars a tourist can harpoon a white beluga whale in Canada's Hudson Bay from a canoe outfitted with an outboard engine. Other members of the marine mammal family are also undergoing the tribulations of civilization. The U.S. Navy utilizes San Miguel Island as a "danger zone," that is, a punching bag for missile salvos. This small isolated island off southern California happens to be one of the last major American haunts of the sea elephant, a two-ton edition of a seal. Like the gray whale, the sea elephant was once hunted to near extinction. Like the gray whale, the sea elephant recovered from the "vanishing" category through prohibitions on commercial kiling. Like the gray whale, the sea elephant now faces eviction from a vital breeding ground. (The Union Oil blowout in the Santa Barbara Channel doused part of the island rookey grounds; investigators found 45 oil-smeared seals on the verge of death.)

The threat of such evictions will occur with increasing frequency until international forms of protection effectively preserve the living space as well as the life stocks of marine mammals. Such agreements have been evolved to protect migratory waterfowl. Because of the whale's oceanwide migration habits, its protection will require the cooperation of nations at all levels of economic development and conservation management. This is a large order, but without such organized foresight whales will follow the buffalo, the mountain gorilla, and the bald eagle to extinction. It seems inconceivable, but so is the very existence of a whale — even when you see one emerge from the quiet waters of Scammon Lagoon in a gallant spy hop.

—Wesley Marx (1967)

22 *A California gray whale "spy hops" at Scammon Lagoon. (Photo courtesy of NOAA)*

The Whale

John Donne

At every stroke his brazen fins do take,
More circles in the broken sea they make
Than cannons' voices, when the air they tear:
His ribs are pillars, and his high arch'd roof
Of bark that blunts best steel, is
thunderproof:
Swim in him swallow'd dolphins, without fear,
And feel no sides, as if his vast womb were
Some inland sea, and ever as he went
He spouted rivers up, as if he meant
To join our seas with seas above the
firmament.

He hunts not fish, but as an officer,
Stays in his court, at his own net, and there
All suitors of all sorts themselves enthrall;
So on his back lies this whale wantoning,
And in his gulf-like throat sucks every thing
That passeth near. Fish chaseth fish, and all,
Flyer and follower, in this whirlpool fall;
O might not states of more equality
Consist? and is it of necessity
That thousand guiltless smalls, to make one
great, must die?

Now drinks he up seas, and he eats up flocks,
He jostles islands, and he shakes firm rocks.
Now in a roomful house this Soul doth float,
And like a prince she sends her faculties
To all her limbs, distant as provinces.
The sun hath twenty times both Crab and
Goat
Parched, since first launched forth this living
boat;
'Tis greatest now, and to destruction
Nearest; there's no pause at perfection;
Greatness a period hath, but hath no
station. . . .

— John Donne

An Open Letter to the Prime Minister of Japan

Advertisement by the Whale Protection Fund, coinciding with the Spring 1977 visit of Japanese Prime Minister Fukuda to Washington.

WHY DOES JAPAN CONTINUE COMMERCIAL WHALING?

This year 32,000 whales will be killed by explosive harpoons—more than 30% at the hands of Japanese whalers. It is a needless slaughter that has driven the great whales, among the most intelligent and highly-evolved animals, to the brink of extinction.

Although all eight species of great whales are on the U.S. endangered species list, Japan had repeatedly ignored the appeals of the United Nations, as well as scientists and conservationists, for a ten-year moratorium on commercial whaling. It is time for Japan to act responsibly in the world community, and to end its short-sighted destruction of marine life.

The history of whaling is a story of overexploitation of species after species. Over the past half-century, more than 2,000,000 whales have died by the harpoon. The blue whale, the largest animal ever to live on Earth, has been virtually exterminated. In one year alone, more than 30,000 blue whales were killed; Japan only stopped blue whaling in 1966 after its fleets *could not find a single blue whale to kill.* Today, only a few thousand blue whales survive, spread across the vast seas. The blue whale may never be able to regenerate.

When the blue whale became commercially extinct, the whalers turned their harpoons on each successively smaller species. Now they are methodically slaughtering the minke whale, a species so small (just one-twentieth the size of a blue whale) that it was not considered commercially valuable until just a few years ago.

The seas surrounding Antarctica are the last refuge for large numbers of whales. Everywhere else they have been reduced to remnant populations. But even in the Antarctic, they are disappearing rapidly.

WHALING IS A DYING INDUSTRY

Japan's Antarctic catch last year was just 5% of the peak year of 1964. One of the Japanese whaling fleets lost more than $5,000,000 because there were so few whales. Employment has plunged to less than 1500 workers in the entire Japanese whaling industry. And last year, all the Japanese whaling companies were merged to form the Japan Joint Whaling Company.

An Open Letter to the Prime Minister of Japan
Advertisement by the Whale Protection Fund (Center for Environmental Education) in the *Washington Post,* March 21, 1977, coinciding with the visit of Japanese Prime Minister Fukuda to Washington, By courtesy of the Center for Environmental Education.

The economics of whaling are so poor that the Soviets, who kill more than 40% of the whales, recently proposed to combine their whaling operations with the Japanese in the Antarctic. Imagine: on the *last* day of the *last* whaling season, Japanese and Soviet gunners blast their harpoons into the *last* whale.

Prime Minister Fukuda, *why does Japan allow the senseless slaughter of whales to continue? What will the future generations of mankind say when they read that Japan condoned the destruction of the great whales in order to provide employment for a few hundred workers, for a few more years?*

There is no real need for whale products in Japan or around the world. The slaughtered whales are converted into chicken feed, margarine, comestics and lubricating oil. There are inexpensive and plentiful substitute sources for all these products.

The Japanese whaling industry claims whale meat is important to the Japanese diet. Yet Japan, the only nation to consume appreciable quantities of whale meat, get much less than 1% of its protein from whale meat. In further irony, Japan kills thousands of sperm whales, whose meat is inedible.

A boycott against the products of Japan and the Soviet Union was launched in 1973 by the major conservation, environmental and animal welfare organizations of the United States. Since that time, millions and millions of Americans have joined the boycott as an act of conscience in response to the Japanese and Soviet refusal to adopt the ten-year moratorium called for by the United Nations. No nation has the right to exterminate life on earth in the name of profit. Whales live in international waters. They are the heritage of all mankind.

WHY DOES JAPAN SUBVERT WHALE CONSERVATION?

Each year, the International Whaling Commision sets whale quotas in order to protect the most endangered species. Japan rejected the 1974 whale quotas, killing thousands more. Stung by the resulting international outcry, then-Prime Minister Takeo Miki pledged in 1975 at a Washington press conference, *". . . The Japanese Government is going to abide by, without any reservations, the conclusions reached by the International Whaling Commisssion. I am well aware of the need for preservation of the whales as a mammal and therefore we are going to abide by any responsible scientific research findings."*

Unfortunately for the whales, the Japanese government tolerates an illicit trade in whale products by encouraging the whaling efforts of several whaling nations who have refused to join the IWC and abide by its quotas. These nations kill thousands of endangered whales outside any regulation. Most of these outlaw whalers are owned or supported by the Japanese whaling industry.

In Peru, for example, the whaling station is owned by the Nippon Hogei Co. Here, more than 1800 whales are killed annually, many of them undersize. According to whale scientists, whale populations on the west coast of South America are being severely depleted by the Peruvian whalers, who ship more than 2500 tons of whale meat to Japan each year.

Japan also makes the market for whale meat from South Korea, where whaling stations take hundreds of whales, including dozens of fin whales, a species under total IWC protection in the Pacific Ocean.

In February, 1977] the U.S. Marine Mammal Commission recommended that President Carter embargo the fishery products of Peru and South Korea under authority of the Pelly Amendment to the Fishermen's Protective Act. The unregulated whaling and refusal to join the IWC by Peru and South Korea trigger the sanctions. But Japan is equally at fault.

Japan's most flagrant subversion of whale conservation is its continued purchase of meat from the pirate whaling ship *Sierra.* A combination factory ship and catcher boat, the *Sierra* was thrown out of the Bahamas several years ago after it nearly wiped out the humpback whale population in the Caribbean. Since then, the ship has been operating off the west coast of Southern Africa, an area declared totally off limits by the whaling commission.

The *Sierra* takes more than 450 endangered sei and Bryde's whales annually, and there are fears it has also been killing the extremely rare southern right whale. Four Japanese meat merchants are aboard at all times to select the best cuts of meat from the whale carcasses. Most of the remaining meat and all the blubber and bones are thrown overboard to the sharks. What remains is deceptively marked "Product of Spain" and shipped to Japan via the Ivory Coast.

We don't know who owns the *Sierra.* A dummy corporation in Lichtenstein is the registered owner, the ship flies a Somali flag of convenience, the captain is Norwegian and the crew is South African. It is obvious, however, that Japan supports this rogue pirate whaler by buying its catch.

When the United States introduced a resolution at last year's IWC meeting calling for all member nations to prohibit the importation of whale products from non-IWC nations, Japan angrily attacked the move, claiming it would interfere with "free trade." Japan led the successful effort to block the resolution.

In another attempt to circumvent whaling regualtions, Japan shocked the conservation and scientific communities last fall by unilaterally announcing it was taking out a "scientific permit" to kill 240 Bryde's whales during the 1976-77 Antarctic whaling season. The IWC has set a zero quota on the killing of this whale population, but Japan took advantage of an obscure loophole in IWC rules allowing limited taking of whales for "scientific study." But Japan gave no justification for its huge kill. Indeed, marine biologists insist there is no scientific justification. The real Japanese motive was clear when they announced that "the whales will not be wasted" and "will be made available to a factory ship for processing."

It is incredible that the Japanese government allows the whaling industry to dictate such policy in the face of the international outcry against whaling. Moreover, the Japanese government is insensitive to the demands of the international community calling for a complete halt to whaling around the world.

In 1972, the 53 nations gathered in Stockholm for the United Nations Conference on the Human Environment voted unanimously for a ten-year moratorium on commercial whaling. Twice since then, the U.N. has unanimously reaffirmed that appeal. Still, Japan and a handful of other nations mercilessly hunt these vanishing species.

Whales are awesome, extraordinary creatures. Their haunting songs have captivated millions of humans. We have only begun to learn about these gentle giants. The whales are the highest form of life in the sea, but they are being systematically destroyed by the highest terrestrial creature . . . humans.

Jacques Cousteau observes sadly: *"The only creatures on Earth that have bigger—and maybe better—brains than humans are the Cetacea, the whales and dolphins. Perhaps they could one day tell us something important, but it is unlikely that we will hear it. Because we are coldly, efficiently and economically killing them off:*

Prime Minister Fukuda, the American people urgently appeal to you to assume strong leadership in halting the slaughter of the great whales by"

1) Ending whaling by Japan;
2) Forbidding the importation into Japan of whale products from other whaling nations;
3) Using Japan's influence as a world power to persuade all nations to end commercial whaling.

なぜ大勢のアメリカ人は日本とソ連の製品を買うことを拒むのでしょうか?

そして、どうしたら貴方はこのボイコットを中止させることが出来るでしょうか。

多数のアメリカ人は日本やソ連で出来た多くの立派な製品のボイコットを支持しています。なぜでしょうか? それは、これら両国が捕鯨を続ける政策をとっているのに抗議するためです。私達は、もし皆さんが次の諸事実を読めば、皆さんは政府の捕鯨政策をかえるようにアピールすることを望まれると思います。

一、大規模な商業ベースでの捕鯨に従事しているのは、日本とソ連の二国だけです。(二国の分を合せると、全世界の鯨の捕獲量のほとんど八五パーセントにも及びます)

二、一九七二年の国連環境会議は、世界各国に対して十年間の捕鯨停止を勧告しました。しかし日本とソ連は生存している鯨を救っていこうという国際的なアピールを無視しています。

三、捕鯨産業は縮少化の傾向にあります。捕鯨にいまなお従事している多数の日本の方々は、より良き拡張しつつある産業に吸収され得る筈です。

四、捕鯨はいかなる重要な経済的目的をも果しません。鯨肉は世界の食糧供給のごく小さな一部でしかありません。(日本においてすら、これは総蛋白質の消費量の一パーセント以下です)

石油製品が家庭用、工業用の鯨油にとって代っています。鯨骨に代る合成代用品はより能率よく、より安く人類の役に立っています。

五、大型鯨の十種類の中、八種類までは絶滅の憂目に頻しています。鯨を殺すことを続けることは不必要なだけでなく、さらに世界からこれらのすばらしい海洋のジャイアントを奪い去ります。

世界は、日本が商業的捕鯨を止めるリーダーシップをとってくれることを期待しています。どうぞ

東京都千代田区永田町
一ノ六ノ一
福田総理大臣様

あてに一筆したためて、貴方は捕鯨は不必要であり、停止すべきであると信ずる、と知らせてください。世界における日本の環境保護者としてのイメージを回復する一助としてください。経済的ボイコットをやめさせるようにしてください。そして、鯨を保護するようご助力ください。鯨を救うことの世界的重要性についてもっと情報のほしい方は右記にご連絡ください。

The Whale Protection Fund
c/o Center for Environmental Education
2100 M Street, Northwest
Washington DC 20037
U.S.A.

An Appeal to the Japanese People

The following is copy from an ad placed in Shukan Asahi *a Japanese news weekly, in April 1977, in an effort to explain the proposed boycott of Japanese products.*

Why Many Americans Are Refusing to Buy Japanese and Soviet Products. . . and how you can help to stop the boycott. Millions of Americans are supporting a boycott of the many fine products manufactured by Japan and the Soviet Union. Why? As a protest to the continued whaling policy which both nations endorse. We believe that when you read the following facts, you will want to appeal to your government to change its pro-whaling policy:

"1. Japan and the Soviet Union are the only two nations involved in large-scale commercial whaling. . . ;

"2. The 1972 UN Environment Conference recommended to *all* nations a 10-year moratorium on all whale killing; yet, Japan and the Soviet Union have continued to ignore the international call to save the remaining whales;

"3. Whaling is already a diminishing industry. . . ;

"4. Whale killing serves no major economic purpose. Whale meat is only a very, very tiny part of the world's food supply. . . ;

"5. Eight of the ten species of great whales are threatened with extinction. Continued slaughter of whales is not only unnecessary, it also threatens to deprive the world of these marvelous giants of the sea.

"All the world is looking to Japan to assume the leadership in ending commercial whaling.... Please help restore Japan's conservationist image around the world. Please help to end the economic boycott. Please help to save the whales . . . . "

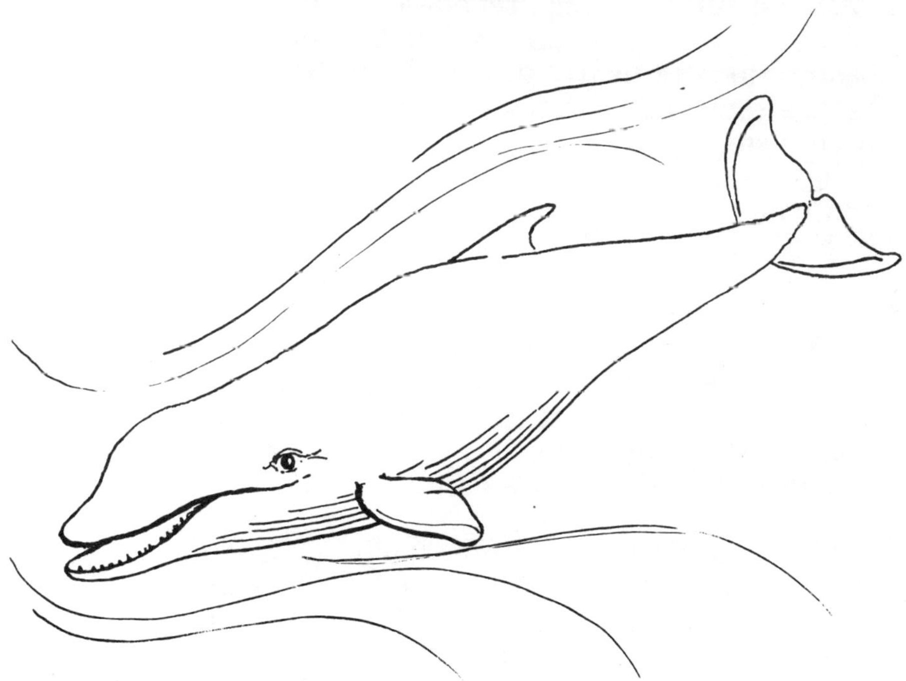

An Appeal to the Japanese People
Courtesy, Center for Environmental Education.

Greenpeace: Making "War" to Save the Whales

Peter Gorner

"Drown, you bloody bastards, drown!"
—Typical reaction to Greenpeace

Driven to obsession by the "ma-ma-ma-ma" cries of terrified baby seals, he has bobbed on Atlantic ice floes to confront their killers. Amid the bloody spume and agonized thrashings of dying whales, he has clung to his puny rubber raft deep in the churning Pacific calmly defying grenade-tipped harpoons. Paul Watson, 26, has been shot at, beaten up, and nearly drowned and frozen as he has repeatedly protested human barbarism.

In a drama of *deja vu,* Watson and 27 men and women crew members will depart Sunday from Honolulu in their 170-foot refurbished World War II subchaser to scour the Pacific for the elusive Japanese whaling fleet. Once located, the Canadian-Hawaiian volunteers of Greenpeace VIII will launch 12-foot rubber zodiac rafts and streak between the catcher boats and the magnificent sperm whales they intend to kill, daring the whalers to kill men, instead.

"We call it aggressive nonviolence," Watson says. "If they want to fire their harpoons at us, they may. We cannot morally get out of the way. If they decide to kill us, we cannot save ourselves. Because then we would mean nothing, then we're just a bunch of words."

Greenpeace. The bucolic name belies the radical (some say suicidal) activism of the fastest-growing environmental group in the world. On whale cruises in 1975 and 1976, the Greenpeace band personally rescued 100 whales. But the threat of Greenpeace publicity is credited with saving more than 1,300 of the giant cetaceans and has been the most dramatic curb against the $200 million whaling industry.

More than 12,000 people have formed 35 Greenpeace chapters in the United States, Canada, and a dozen foreign countries. Scientists, technologists, lawyers, doctors, and concerned persons have joined the intrepid kids from Vancouver, B.C., who in 1970 rented an 85-foot halibut seiner and steamed toward Amchitka Island, planning to plop anchor there to oppose United States atomic testing in the Aleutians.

"Our roots were Quaker, but most of us were conservationists frustrated by the stodginess of the Sierra Club." says Watson, a native of New Brunswick. "The U.S. Coast Guard grabbed the Greenpeace I. I was on Greenpeace II, the relief vessel, but we still were 100 miles off Amchitka when they exploded the bomb one day early. That blast killed more than 1,000 sea otters, plus countless seals and other marine mammals. But we had made our point. Five other tests were canceled."

Greenpeace: Making "War" to Save the Whales
Reprinted courtesy of the *Chicago Tribune,* June 16, 1977.

Watson, interviewed at a wildlife conference in St. Louis, contends the nonprofit Greenpeace Foundation always is broke. Still, its members scrounged up enough cash to sail a ship into the French nuclear test zone in the South Pacific in 1972 and 1973. The first time, a French navy destroyer rammed Greenpeace III, disabling it. The following year, French sailors boarded the Greenpeace vessel and assaulted its captain, Vancouver yachtsman David McTaggart, blinding him in one eye.

"David sued and the facts of international piracy weren't disputed," Watson says. "But the case has dragged on for years. We place little faith in courts. The only thing that works is direct confrontation. All the understanding, the government commissions, the studies don't stop reckless people. Creating international incidents does."

In 1975, Greenpeace turned to whales. So did much of the world. Whale lore and oceanic ecology swept across college campuses. In the last 50 years 200 million whales have been slaughtered. Every 14 minutes, another whale is killed. In 1971, whaling and whale products were banned in the U.S. The Marine Mammal Protection Act 1972 prohibits any American from taking or harassing whales. But 85 per cent of the whaling business is conducted by Japan and the Soviet Union Japan only stopped killing blue whales when it coulden't find any more.

The International Whaling Commission, the industry's self-regulating body, has just gone into seclusion to pound out this season's quotas. The giant blues have been protected since the 1960's — in 1963, the commission reduced its quotas to 15,000 blue whales, but whalers could only find 112. California's gray whales are protected as they meander their way 12,000 miles from Mexico to Alaska and back. But the humpback, right, bowhead, fin, sei, Bryde's, sperm, and minke whales may soon face extinction.

The whale has become the symbol of conservation. Fascinated buffs have listened to the deep basso humming, moaning, and gurgling of the humpbacked whales, down to 7 per cent of their original 100,000. Students have learned the differences between the toothed whales and baleens like the blue, which convert the most elemental sea animals (eight tons of plankton a day) to bulk protein.

School children learn that a 200-ton blue whale weighs as much as 3,000 of them. His tongue dwarfs a Volkswagen. Thirty-three elephants could fit inside his skin. A trout could swim in his arteries. But the formidable blues really are intelligent, peaceful, graceful, sometimes playful, and always beautiful. There once were 200,000 blue whales. Now they are nearly gone.

The animal the hunters really want is the sperm whale, *Physeter catodon,* the 65-foot-long, bluish-gray, sharp-toothed leviathan with the oversized head that holds reservoirs of oil. Whalers use airplanes to spot the close-knit harem herds of 50 or more, with playful bulls blowing clouds of steam in that characteristic 45 degree angle.

Sperm whales are capable of diving a mile or more to challenge 60-foot-long giant squids, their favorite meal. When enraged, sperms have rammed and sunk ships. Yet they nobly aid wounded comrades, sometimes tremble at the threat of something unknown, rush away in terror, even become so paralyzed they let themselves be killed. The sperm is a sensitive animal far more affected by moods of its fellows than most other mammals. Watson has been among them at his most vulnerable — when they were in their greatest agony. Yet he never has been hurt.

"Once they're spotted by air, it's nearly over," he says. "Fast boats chase them ruthlessly as they surface, sonar tracks them when they dive. Finally, the whales are exhausted. Then they (hunters) fire a harpoon and it explodes deep inside the whale. The cascades of blood, the gore, the screams — no one who has witnessed the death of such a creature is unaffected. We have many ex-whalers among our membership."

Watson is in charge of the confrontation crews. Greenpeace volunteers first spend weeks learning basic seamanship, how to tie knots, communications, navigation. A select team works with the 35 horsepower-powered rubber zodiacs. Watson wants his people to be able to touch the rear pontoon of the zodiac to the bow of a ship moving at 16 knots.

On its first cruise, Greenpeace had no trouble locating the Soviet fleet, but the Russians simply outran them. The next year, they had received enough donations to lease a faster ship.

"We located the Russians 60 miles due west of Eureka, Cal.," Watson says. "After launching the zodiacs, we managed to save eight whales by blocking them to escape. Our zodiacs were about a mile from our mother ship. The harpooner kept kneeling and bobbing, trying to get a shot off, but we were almost riding on the backs of the whales.

"Then a wave picked up a zodiac and tossed it on its side. The Russian captain made a throat-slashing gesture and tried to run us down. We formed a line of zodiacs, 20 yards in front of the whale. Then the harpoon went over my head, the rope missed us by about four feet. But the 250-pound harpoon struck the whale. It was a female. The enraged male charged the Russian boat. Then they shot him, too.

"When they loaded the whales onto the processing ship, we rammed our zodiacs into the slipway. That delayed them. But they turned the fire hoses on us. The pictures, though, went around the world."

The resulting publicity had meat. Last summer, Greenpeace leased a 153-foot former minesweeper, which was renamed Greenpeace VII. The group fixed it up for $40,000 in exchange for use. But once located, the Russians merely quit firing and backed off. Zodiac members exchanged gifts with harpooners — cigarets and whisky for Greenpeace buttons and T-shirts.

The crew did have anxious moments when radio contact was lost to two zodiacs. They returned unharmed, but it points to the seriousness of the cause. No Greenpeace member is paid (on the recent campaign for baby harp seals, the foundation spent $60,000 and still owes $20,000) and many crew members quit jobs and rearrange lives.

Last summer, the crew included a Russian-speaking captain and Japanese nurse, both of whom intercepted radio signals from the whaling fleets.

"It will be difficult to find the Japanese this time," Watson says. "They never use the word 'whale' over the radio. They just say 'fish.' So we'll have to check out dozens of fishing boats until we find whalers. But we got some tricks of our own planned for them."

Watson fears time is running out, both for endangered animals and for his type of nonviolence. "I think a Greenwar, an eco-war, is inevitable unless something changes the thinking of business and government. We Greenpeace people won't do it, but we're constantly being approached with offers. People want to give us bombs and guns. People want to see Japanese and Russian whaling ships and Norwegian sealers blown up.

"Yet I think we've shown that a small group of 30 people — and it's never been more than that — can actively accomplish something. And people either love us or hate us for it."

—Peter Gorner, (1977)

23 A gigantic sperm whale lies dead on shore. (Photo courtesy of NOAA)

Stranded High and Dry

Not all the dangers that whales face come from humans (although those are the most threatening to the species as a whole). From time to time, ever since people have observed whales near their shores, a few unlucky creatures have been stranded. New Englanders began their whaling trade with this sort of shore whaling. Scientists have not yet agreed on the explanation for the almost perverse way in which whole groups of whales not only strand themselves but persist in swimming back to the land rather than out to safety. Besides the incident reported here, there have been other mass beachings in Florida—a group of 60 pilot whales came ashore on Grassy Key in 1966, and, in 1970, a group of 150 false killer whales beached themselves near Fort Pierce. . . . Why?

Although whale beaching is not a totally uncommon phenomenon along the Florida coast, the incidents of mass beachings by spinner dolphins and false killer whales in the summer of 1976 aroused the curiosity of many marine biologists and concerned citizens.

The first incident occurred on July 14th when a group of 25 spinner dolphins washed ashore and died along a beach near Sarasota. The next incident occurred one week later when five false killer whales beached near Fort Myers, some 50 miles south of Sarasota. Rescuers were able to save four of the whales and transport them to a nearby marine aquarium, Sea World. Of the four, one has subsequently died. In captivity, false killer whales have only a 95 percent chance of surviving. From aerial photographs taken at the time of the beaching, there are indications that as many as 25 other whales may have entered the same area through one pass close to the shallow water, but then escaped through another pass.

The third beaching occurred on July 26th when 30 false killer whales stranded themselves on Loggerhead Key, a tiny island in the Dry Tortuga chain, approximately 150 miles southwest of Fort Myers. By now, biologists from the University of Miami, University of Florida, and the National Marine Fisheries Service, as well as experts from the Florida Marine Patrol, the U.S. Coast Guard, and the Miami Seaquarium, were well alerted to the growing number of beaching incidents. And they were prepared to embark upon a major rescue effort, in an attempt to avoid the tragic losses resulting from earlier beachings.

After several vain attempts, the rescue team decided their only alternative was to literally "drive" the animals out into deeper waters. Already one of the whales had died from exposure. Whales are extremely social mammals, traveling in groups or "pods," with normally the largest whale as the leader. Initial reports suggested that the dead whale was the largest, and that possibly the other, more healthy whales in the pod had instinctively followed the sick leader onto the

---

Stranded High and Dry
From *Conservation News,* September 15, 1976, courtesy of National Wildlife Federation.

beach. However, biologists aren't certain that this is a totally accurate interpretation.

During the night prior to the final rescue attempt, the animals were doused with water to prevent dehydration. Early the next morning, the rescue team tagged, measured, took blood samples and photographed each of the surviving whales. It was later through these photographs that Dr. Daniel Odell, a University of Miami biologist and member of the rescue team, was able to determine that several of these whales were among those which had approached the shallow water in the Fort Myers incident one week earlier. He made this determination through comparison of pictures of the whales' dorsal fins, which in whale identification, Dr. Odell claims, can be used "almost like fingerprints."

With the rescue operation underway, team members began to drive the animals offshore by swimming alongside, pushing and riding the whales. Three rescue boats were used to corral the animals until they were about one mile from Loggerhead Key. Once they were out far enough, the "corral" was opened, and the whales returned to their Atlantic home.

The whales being successfully rescued, biologists now returned to the task of analyzing autopsies of the dead dolphins and whales, in an attempt to determine what caused the beachings. In *all* of the dead mammals parasite infestations were present. And in the whales, these parasites had penetrated into the middle and inner ear sinuses. Although the parasite infestation remains only a theory as to the cause of the beachings, scientists believe it could have affected the whale's sense of direction and balance enough to make it run aground. In one report, this situation was likened to a ship with a radar malfunction.

No sooner had these findings been made, when seventeen more dead false killer whales were discovered in shallow waters off Cape Sable in Everglades National Park. It is possible that the seventeen were part of the 29 that had been rescued off Loggerhead Key, but the carcasses were so badly mauled by sharks that no accurate photographic comparison could be made.

According to Dr. Odell the reason for the beachings is not really clear. Although the parasite infestation is perhaps the most reasonable explanation, there are some flaws in this theory. The presence of parasites is quite normal among most wild animals. They are naturally ingested through food sources. The real question, then, is what kind of parasite load is normal for these marine mammals. For if this load was not exceeded, it is doubtful that the infestations could have caused the beachings....

This most recent surge of beachings does not necessarily mean that a rising trend is being established. Mass beachings have always plagued Florida. In an incident in 1970, 150 false killer whales stranded themselves, and in 1971, a large group of pilot whales were found beached along Florida's eastern coast. The difference is that today many of the beachings, which might have gone unnoticed in the past, are now being reported, because, according to Dr. Odell, "there is more of an awareness by the public in general. . . ."

—National Wildlife Federation, (1976)

24 Once inhabited by Wampanoag Indians, Nantucket Island, Massachusetts, is now a summer tourist town with restaurants, shops, bright flowers, and low, picturesque cottages bordered by little white fences. (Photo courtesy of Massachusetts Department of Commerce and Development)

4

TALES OF WHALES

Oh, the rare old Whale,
mid storm and gale
In his ocean home will be,
A giant in might,
where
might is right,
And King of the boundless sea.

—Whaling song

The Adventures of Jonah

Jonah rose to flee to Tarshish from the presence of the Lord. He went down to Joppa and found a ship going to Tarshish; so he paid the fare, and went on board, to go with them to Tarshish, away from the presence of the Lord.

But the Lord hurled a great wind upon the sea, and there was a mighty tempest on the sea, so that the ship threatened to break up. Then the mariners were afraid, and each cried to his god, and they threw the wares that were in the ship into the sea, to lighten it for them.

But Jonah had gone down into the inner part of the ship and had lain down, and was fast asleep. So the captain came and said to him, "What do you mean, you sleeper? Arise, call upon your god! Perhaps the god will give a thought to us, that we do not perish."

And they said to one another, "Come, let us cast lots that we may know on whose account this evil has come upon us." So they cast lots, and the lot fell upon Jonah.

They said to him, "Tell us, on whose account this evil has come upon us, what is your occupation? And whence do you come? What is your country? And of what people are you?"

He said to them, "I am a Hebrew; and I fear the Lord, the God of heaven, who made the sea and the dry land."

Then the men were exceedingly afraid and said to him, "What is this that you have done!" For the men knew that he was fleeing from the presence of the Lord, because he had told them.

Then they said to Jonah, "What shall we do to you, that the sea may quiet down for us?" For the sea grew more and more tempestuous.

Jonah answered, "Take me up and throw me into the sea; then the sea will quiet down for you, for I know it is because of me that this great tempest has come upon you."

Nevertheless, the men rowed hard to bring the ship back to land, but they could not, for the sea grew more and more tempestuous against them. Therefore they cried to the Lord, "We beseech thee, Of Lord, let us not perish for this man's life, and lay not on us innocent blood, for thou, O Lord, has done as it pleased thee!"

So they took up Jonah and threw him into the sea, and the sea ceased from its raging.

Then the men feared the Lord exceedingly, and they offered a sacrifice to the Lord, and made vows.

And the Lord appointed a great fish to swallow up Jonah. And Jonah was in the belly of the fish three days and three nights.

Then Jonah prayed to the Lord his God from the belly of the fish, saying:

"I called to the Lord, out of my distress,
And he answered me.
Out of the belly of Sheol I cried,
And thou didst hear my voice.

For thou didst cast me into the deep, into the heart of the sea,
And the flood was round about me;
All thy waves and thy billows passed over me. . . .
The waters closed in over me,
The deep was round about me;
Weeds were wrapped about my head at the roots of the mountains.
I went down to the land whose bars closed upon me forever. . . .
But I with the voice of thanksigiving will sacrifice to thee;
What I have vowed I will pay.
Deliverance belongs to the Lord!"
And the Lord spoke to the fish, and it vomited out Jonah upon the dry land.

Whale Watchers: Hamlet, Prince of Denmark

Hamlet: Do you see yonder cloud that's almost in shape of a camel?
Polonius: By the mass, and 'tis like a camel, indeed.
Hamlet: Methinks it is like a weasel.
Polonius: It is backed like a weasel.
Hamlet: Or like a whale?
Polonius: Very like a whale.

—William Shakespeare

From *Hamlet,* by William Shakespeare.

The Flying Whaleman

Herman Melville

I heard of one whaler, which after many years' absence was given up for lost. The last that had been heard of her was a shadowy report of her having touched at some of those unstable islands in the far Pacific, whose eccentric wanderings are carefully noted in each new edition of the South Sea charts. After a long interval, however, the *Perseverance*— for that was her name—was spoken somewhere in the vicinity of the ends of the earth, cruising along as leisurely as ever, her sails all bepatched and bequilted with rope-yarns, her spars fished with old pipe staves, and her rigging knotted and spliced in ever possible direction. Her crew was composed of some twenty venerable Greenwich-pensioner-looking old salts, who just managed to hobble about deck. The ends of all the running ropes, with the exception of the signal halyards and poop-down-haul, were rove through snatch-blocks, and led to the capstan or windlass, so that not a yard was braced or a sail set without the assistance of machinery.

Her hull was encrusted with barnacles, which completely encased her. Three pet sharks followed in her wake, and every day came alongside to regale themsleves from the contents of the cook's bucket, which were pitched over to them. A vast shoal of bonetas and albicores always kept her company.

Such was the account I heard of this vessel, and the remembrance of it always haunted me. What eventually became of her I never learned; at any rate she never reached home, and I suppose she is still reguarly tacking twice in the twenty-four hours somewhere off Buggerry Island, or the Devil's-Tail Peck.

—Herman Melville
Typee

An Interview With King Cachalot

Frank T. Bullen

A working whaleman for many years, the author turned to writing in the late 1890s and produced the adventure classic, The Cruise of the Cachalot *, as well as many short sketches. He seems to have been unique among whaling writers in his ability to anthropomorphize the whale, both humorously (as in this sketch) and pathetically, as in sketches like "The Loss of the First-Born." London-born, Bullen constantly parodied the Yankee accent.*

Difficulties, which, could I have foreseen then, would have appeared insurmountable, attended the interview hereinafter recorded. First of all, His Majesty King Cachalot the MMCC was not in the best of humours — which was hardly to be wondered at, since, with all the ability we could muster, five boats' crews of us from the spouter *Finback* had been harassing him since daylight, eager to add his fourteen-ton overcoat to our greasy cargo. It was a blazing day on the Line, Pacific side, with hardly a ripple on the water, so that what advantage there was weighed on our side. Yet so wary and skilful had his Majesty proved, that one by one the boats had retired hurt from the field, while the object of their attentions was as fresh as paint, and, as he afterwards expressed it,"going very strong." Nevertheless the scrum had been warm in a double sense, and his Majesty bore many palpable evidences of our efforts all over his huge black body.

Being in command of the only surviving boat, sole representative of our available force, and with a reputation yet to win, I must confess to a little lack of care, a nervous desire to distinguish myself; but I still think it was hard to have my boat knocked into a litter of barrel staves by the unanticipated somersault of my expected prize just as I reckoned upon delivering a *coup de lance* in final settlement of our little account.

After the surprise of our meeting had somewhat subsided, I found myself reclining in a richly carved and upholstered chair in my genial host's splendidly furnished reception room, puffing with appreciative enjoyment at one of his unapproachable Rothschilds — beg pardon I'm sure — I mean that I found myself clinging with no uncertain clutch to a capsized line-tub, into which I succeeded in getting after a series of involuntary evolutions, after having managed to swallow the majority of a barrel of salt water. While settling myself in my ark like a faded Moses, our late antagonist drew near and watched me closely. As soon as I appeared to be *compos mentis* he thus addressed me:

"What you settin' there fur a-gappin' at *me* "sif y' didn't know who i wuz."

"I humbly beg your Majesty's pardon; I meant no offence, I assure you. But I perceive you are an American citizen."

"Perseev' nothin', y'abbrevyated galoot,"

growled he. "Hain't enny persepshun 'baout ye, 'r y'ewd see I' waitin' ter be interviewed, same's all th' other sellebritiz."

Now, although I *do* believe that the journalist is *nascitur, non fit* my nascent journalism if existent was decidedly latent, and at present I was indubitably unfit for anything but a rescue or two. But here was a unique chance of becoming famous, and though modest and retiring to the last degree, I rose to the occasion. A few fragmentary recollections marshalled themselves, and I asked insinuatingly:

"How old is your Majesty?"

"One thousand, four hundred seasons, " he replied promptly.

As soon as I recovered my breath, I answered politely, "Indeed! Your Majesty wears well. I should hardly have thought it. Are your Majesty's parents living?"

"How'd I know," he grumbled, peeping fiercely at me out of the corner of his starboard eye. "Don't go much on parents ermong our peepul. Next please!"

"Where did your venerability do us the honour to be born, if the question be allowable?" I queried timidly.

"Here," he roared, with a resounding crash of his enormous tail on the surface; "wher'd ye think I'd be born but at sea?"

Deficient in locality evidently, I thought, being a bit of a phrenologist myself, though it would have required a theodolite to survey the bumps upon his capacious cranium. But as he showed signs of irritability, I added quickly, "Are you married, your Majesty, or how?"

"Well; I should cackle," he said — "married, hay! Why one of your (an awful reverberation suggested a powerful adjective) slush-tubs hez jest broke up one uv the purtiest little harems I ever collected, twelve ravishin' beauties sech wuz any monark'd be proud of. Well thar, hurry up; I'm jest reminded ov an ole schoolmate uv mine's got mose's good erwun. He's usin' roun' the Bonins 'baout now, 'n' I mus' git over thar'n' b' reave him. Royal rights, y'know," and his Majesty shed a ponderous wink.

"What does your Majesty do for a living?" I ventured to inquire.

"Eat!" he roared. "Harpoons en bomb-guns, what dz ennybody du fr a livin'? I never heerd sech a barnacle-headed grampus'n all my fishin'." With that he lifted up his tremendous caput out of water and exposed his Blackwall tunnel of a mouth, as who should remark, "Not much room for other occupation in a whale's life when a gulf like this needs attention."

I suppose I looked a bit preoccupied, for he hastily added, "But I never eat sech insecks ez you be."

"What, never?" I ventured to murmur.

"No, never," he replied; "at least, that is," — but seeing his hesitation, I said I fancied I'd heard a story about a passenger by the name of Jonah down on the Syrian coast a while back. "Oh, well y'know," he muttered apologetically," 'f course accidents will happen, 's the shark said to his brother when he took him in, but I don't reckon thar wuz anythin' to mek a noise erbout. 'Tanyrate the can'date left considerable sudden. Yew needn't be 'fraid ennyhow."

But I was unprepared with any more questions at the moment, the outlook, or inlook rather, being so disconcerting. So I said, "Would your Majesty object to outlining a few of your wonderful experiences for the benefit of landsmen generally. Any information you may choose to give will be regarded as strictly confidential, of course."

"Oh, sartinly," he replied with an alarming area of smile. "Mos' ov 'em hev ben with your dod-gasted tribe. Why yew're tarnally prowlin' erbout tryin' ter get ter wind'ard ov peac'ble fokes I kaint surmise. Still, up till now I've ben equal ter holdin' me own,—keepin' me eend up, ez yew may say. To-day f'rinstance, hey?" I winced under the sarcasm. "But I mind onst daown on the Noo Seelan' coas' towin' five boat-load ov Mowries frum the Solander 'way down eenamost ter the Cambells. They wuz a plucky crowd, f'r they helt on ter me through a blizzard ov hail an' snow lasting twyst az long as I kin stay soundin'. When it gin over they wuz all fruz stiffer'n a lance-pole. My, but gettin' cleer ov em wuz a pull. I hed to soun' at top-gait 'sif I wuz boun' f'r two thousan' fathoms, 'n' suthin' hed ter give. I wuz pretty fat in those days, so their all-fired irons drew. They galled me like sixty, but I was free.

"Then a left-handed-on-both-feet crowd eout ov a French right-whaler tackled me offn the Cape. Mighty big mistake *they* struck—thought I wuz pore ole say-nothin'-ter-nobody Mr. Cetus, they did. 'N', when I milled roun' 'n' cum f'r 'em eend on 'ith er twenty foot smile on me hed! airthqueeks 'n' volcanose! y' sh'd jest er seen 'em flew. Didn' wait to say howdy, jest cut line 'n' vamoosed like 'sif ole Jemmy Smallback wuz after 'em. I wuz thet mad, I'd liketer hev busted up their ole hooker 'n' all, but thet than *Essex* affair gin me sech er swell'd hed I 'lowed it warn't bad reck'nin' ter let her go et that.

"Say, djever see er big squid, big's me?" he queried sharply.

"Yes, your Majesty, I did once. Only once. B-b-b-ay of B-b-bengal," for I was almost moribund.

"Ah, you *hev* seen suthin' then. F'r yew insecks wut live on top don't offen git a chance ter see them critters 'less we bring 'em up f'r the sun ter see haow gaul-darned ugly they air. Wall, one like yew say yew seen tangled erp my fav'rit' wife off Futuna one afternoon. Me an' my harem wuz feedin' at 'bout a thousan' fathom, an' Polly jest sidled up ter ole Jellybelly 'n' got hole ov a mouthful ov him. He, bein' kinder s'prised, gripped her all over ter onst; 'n' stranger," he added impressively, "I'll be weather-bound ef he didn't frap her hole head up so's she couldn't bite er breathe. We'd ben down 'bout long ernough too, but I sailed right in 'n' bit his great carkiss in half az well az I c'd see f'r his ink-cloud. How'sever I wuz too late, f'r he'd locked his tangle ov arms roun' an' roun' her hed, 'n' though his body wuz all chawed erp they couldn' come adrift. So she drowned, 'n we all hed ter make tracks unstairs quicker 'n winkin' er we sh'd a ben drowned tu. As 'twuz we wuz fair beat out when we arrove up top.

"Did I ever have enny fights with me own people? Well I—but there, how'd yew know, poor thing. Millyuns ov'em. Look at me," and he swept proudly past exhibiting his grooved and ribbed flanks bearing indelible traces of many a furious battle, some of the foot-wide scars being twenty feet long.

"Enny more informashun I c'n supply yew with at short notice? bekuz this session'll hev ter adjurn *siny die* in about tew minnits. I'm gittin' mos' amazin' peckish."

Happy thought, "What do you live on mostly, your Majesty?"

"Squid. Fust 'n las' 'n' between meals gen'ly. They aint nothin' better tew eat in the hull worl' 'z far's I know. We dew 'casionally git a bellyful ov fish ov sorts by layin' quiet when the shoals air swarmin'. They run down a feller's gullet in hunderds 'n never know whar they're goin. But they're cussid indigestible—"

I was alone. There was nothing in sight, but my interviewee was gone. So stiff and sore was I that I could hardly turn my head to see if help was coming. There was no help in sight that I could discover, but presently a boat came along from the ship and picked me up—none too soon. Gloomily we returned on board to moralise mournfully over our ill-luck and the perfidy of sperm whales generally.

—Frank T. Bullen,
Idylls of the Sea (1899)

Whale Watchers: Henry David Thoreau

We saw this forenoon what, at a distance, looked like a bleached log with a branch still left on it. It proved to be one of the principal bones of a whale, whose carcass, having been stripped of blubber at sea and cut adrift, had been washed up some months before. It chanced that this was the most conclusive evidence which we met with to prove what the Copenhagen antiquaries assert, that these shores were the *Furdustrandas,* which Thorhall, the companion of Thorfinn during his expedition to Vinland in 1007, sailed past in disgust. It appears that after they had left the Cape and explored the country about Straum-Fiordr (Buzzard's Bay!), Thorhall, who was disappointed at not getting any wine to drink there, determined to sail north again in search of Vinland. Though the antiquaries have given us the original Icelandic, I prefer to quote their translation, since theirs is the only Latin which I know to have been aimed at Cape Cod.

> "Cum parati erant, sublato velo, cecinit Thorhallus: Eo redeamus, ubi conterranei sunt nostri! faciamus aliter, lata navis explorare curricula: dum procellam incitantes gladii morae impatientes, qui terram collaudant, Furdustrandas inhabitant et coquunt balaenas."

In other words, "When they were ready and their sail hoisted, Thorhall sang: Let us return thither where our fellowcountrymen are. Let us make a bird (a vessel) skillful to fly through the heaven of sand (the sea) to explore the broad track of ships; while warriors who impel to the tempest of swords; who praise the land, inhabit Wonder Strands, *and cook whales.*" And so he sailed north past Cape Cod, as the antiquaries say, "and was shipwrecked on to Ireland."

Though once there were more whales cast up here, I think that it was never more wild than now. We do not associate the idea of antiquity with the ocean, nor wonder how it looked a thousand years ago, as we do of the land, for it was equally wild and unfathomable always. The Indians have left no traces on its surface, but it is the same to the civilized man and the savage. The aspect of the shore only has changed. The ocean is a wilderness reaching round the globe, wilder than a Bengal jungle, and fuller of monsters, washing the very wharves of our cities and the gardens of our seaside residences. . . .

—*Cape Cod* (1865)
published

From *Cape Cod* by Henry David Thoreau (published posthumously, 1865).

How the Whale Got His Throat

Rudyard Kipling

Another sort of Jonah story.

IN the sea, once upon a time, O my Best Beloved, there was a Whale, and he ate fishes. He ate the starfish and the garfish, and the crab and the dab, and the plaice and the dace, and the skate and his mate, and the mackereel and the pickereel, and the really truly twirly-whirly eel. All the fishes he could find in all the sea he ate with his mouth — so! Till at last there was only one small fish left in all the sea, and he was a small 'Stute Fish, and he swam a little behind the Whale's right ear, so as to be out of harm's way. Then the Whale stood up on his tail and said, 'I'm hungry,' And the small 'Stute Fish said in a small 'stute voice, "Noble and generous Cetacean, have you ever tasted Man?"

"No," said the Whale. "What is it like?"

"Nice," said the small 'Stute Fish. "Nice but nubbly."

"Then fetch me some," said the Whale, and he made the sea froth up with his tail.

"One at a time is enough," said the 'Stute Fish. "If you swim to latitude Fifty North, longitude Forty West (that is magic), you will find, sitting *on* a raft, *in* the middle of the sea, with nothing on but a pair of blue canvas breeches, a pair of suspenders (you must *not* forget the suspenders, Best Beloved), and a jack-knife, one shipwrecked Mariner, who, it is only fair to tell you, is a man of infinite-resource-and-sagacity."

So the Whale swam and swam to latitude Fifty North, longitude Forty West, as fast as he could swim, and *on* a raft, *in* the middle of the sea, *with* nothing to wear except a pair of blue canvas breeches, a pair of suspenders (you must particularly remember the suspenders, Best Beloved), *and* a jack-knife, he found one single, solitary shipwrecked Mariner, trailing his toes in the water. (He had his mummy's leave to paddle, or else he would never have done it, because he was a man of infinite-resource-and-sagacity.)

Then the Whale opened his mouth back and back and back till it nearly touched his tail, and he swallowed the shipwrecked Mariner, and the raft he was sitting on, and his blue canvas breeches, and the suspenders (which you *must* not forget),*and* the jack-knife — He swallowed them all down into his warm, dark, inside cupboards, and then he smacked his lips — so, and turned round three times on his tail.

But as soon as the Mariner, who was a man of infinite-resource-and-sagacity, found himself truly inside the Whale's warm, dark, inside cupboards, he stumped and he jumped and he thumped and he bumped, and he pranced and he danced, and he banged and he clanged, and he hit and he bit, and he leaped and he creeped, and he prowled and he howled, and he hopped and he dropped, and he cried and he sighed, and he crawled and he bawled, and he stepped and he

lepped, and he danced hornpipes where he shouldn't and the Whale felt most unhappy indeed. (*Have* you forgotten the suspenders?)

So he said to the 'Stute Fish, "This man is very nubbly, and besides he is making me hiccough. What shall I do?"

"Tell him to come out," said the 'Stute Fish.

So the Whale called down his own throat to the shipwrecked Mariner, "Come out and behave yourself. I've got the hiccoughs."

"Nay, nay!" said the Mariner. "Not so, but far otherwise. Take me to my natal-shore and the white-cliffs-of-Albion, and I'll think about it." And he began to dance more than ever.

"You had better take him home," said the 'Stute Fish to the Whale. "I ought to have warned you that he is a man of infinite-resource-and sagacity."

So the Whale swam and swam and swam, with both flippers and his tail, as hard as he could for the hiccoughs; and at last he saw the Mariner's natal-shore and the white-cliffs-of-Albion, and he rushed half-way up the beach, and opened his mouth wide and wide and wide, and said, "Change here for Winchester, Ashuelot, Nashua, Keene, and stations on the *Fitch*-burg Road;" and just as he said "Fitch" the Mariner walked out of his mouth. But while the Whale had been swimming, the Mariner, who was indeed a person of infinite-resource-and-sagacity, had taken his jack-knife and cut up the raft into a little square grating all running criss-cross, and he tied it firm with his suspenders (*now* you know why you were not to forget the suspenders!), and he dragged that grating good and tight into the Whale's throat, and there it stuck! Then he recited the following *Sloka,* which, as you have not heard it, I will now proceed to relate —

By means of a grating
I have stopped your ating.

For the Mariner he was also an Hi-ber-ni-an. And he stepped out on the shingle, and went home to his mother, who had given him leave to trail his toes in the water; and he married and lived happily ever afterward. So did the Whale. But from that day on, the grating in his throat, which he could neither cough up nor swallow down, prevented him eating anything except very very small fish; and that is the reason why whales nowadays never eat men or boys or little girls.

The small 'Stute Fish went and hid himself in mud under the Door-sills of the Equator. He was afraid that the Whale might be angry with him.

The Sailor took the jack-knife home. He was wearing the blue canvas breeches when he walked out on the shingle. The suspenders were left behind, you see, to tie the grating with; and that is the end of *that* tale.

— From *Just-So Stories*
(1902)
Rudyard Kipling

25
Here is the Whale looking for the little 'Stute Fish, who is hiding under the Door-sills of the Equator. The little 'Stute Fish's name was Pingle. He is hiding among the roots of the big seaweed that grows in from of the Doors of the Equator. I have drawn the Doors of the Equator. They are shut. They are always kept shut, because a door ought always to be kept shut. The ropy-thing right across is the Equator itself; and the things that look like rocks are the two giants Moar and Koar, that keep the Equator in order. They drew the shadow-pictures on the doors of the Equator, and they carved all those twisty fishes under the Doors. The beaky-fish are called beaked Dophins; and the other fish with the queer heads are called Hammer-headed Sharks. The Whale never found the little 'Stute Fish till he got over his temper, and then they became good friends again.

—Rudyard Kipling,
Just-So Stories (1902)

Did the Unicorn Have Fins?

Elizabeth B. Simons

The unicorn probably *did* have fins, according to modern scientists.

Everyone knows what a unicorn is supposed to look like. However, few people know that the "unicorn" was probably a tusked whale called the narwhal, which measures 18 to 25 feet in length, weighing up to 2 tons.

In medieval times, rulers afraid of assassination paid huge sums for "unicorn horns" which, when made into cups, were said to be a poison antidote. These "unicorn horns," scientists believe, were the tusks of the narwhal, not of a horned horse.

To catch a unicorn, a maiden would sit in the forest and call the animal to her. According to legend, the unicorn would lay its head in her lap, she would quickly put a golden bridle on it and the unicorn would be in her power. The hunters would then ride up and capture it. However, it is doubtful that "unicorns" were ever caught in any way but with a harpoon. The fabulous tusk is actually an overgrown, twisted (8 to 9 foot) left front tooth, which only male narwhals have. Although no one is sure about the purpose of the whale's tusk, (except the narwhal), people speculate that it is used for defense, courtship, spearing food or making holes in the ice. However, in those cases, females would have a tusk, too.

Narwhals are relatively safe from whalers because of the inaccessability of their Arctic homes. They usually live in groups of about 50, but these groups occasionally combine to form herds of thousands of narwhals. Eskimos do often hunt the narwhal for meat, blubber, hide and tusks. The narwhal hide, when cut into small chunks, is known as *muktuk* and is eaten as a treat. Narwhals feed on cuttlefish, squid and crustaceans.

Remember, the next time you see a unicorn, check to see if it has fins.

— Elizabeth B. Simons

A Tale of Sinbad the Sailor

. . .One day, after weeks of sailing out of sight of land, we saw an island in the sea covered with such fair greenery and trees that it seemed to us to be one of the gardens of Eden. The captain immediately set the ship toward this delightful land, and when the anchor had been dropped and the ship's ladder let down, let his passengers disembark.

All of us merchants landed, carrying with us our food and cooking utensils. The hungry lit their fires and prepared a meal. The fastidious began to do their laundry. The curious and the lazy rested or explored. I was among the last to land and, not neglecting either food or drink, still found time to wander among the trees and take pleasure in the strange vegetation.

All the passengers were occupied in these various ways when the island suddenly trembled from one end to the other, so violently that we were thrown to the ground. As we lay there dazed, we saw the captain leaning over the rail of the ship. He shouted wildly and waved his arms: "Save yourselves! Come aboard! That is no island, but a gigantic whale! She has lived in the middle of the sea since time was young, and trees and shrubs have grown in the sea sand on her back. Your cooking fires have troubled her rest, and now she is moving. Come aboard quickly, for your lives! Or she will dive into the sea and destroy you all."

The merchants jumped up from the ground. Leaving everything behind—clothes, cooking pots, laundry tubs—they rushed toward the ship, which was already weighing anchor. Some of them reached her in time. But the whale, after leaping terrifically two or three times, dived deep into the water, carrying the others beneath monstrous waves.

I was one of those left on the back of the whale, but Allah saved me from drowning by sending a hollow piece of wood drifting toward me. It was a kind of large trough in which some passengers had been washing their clothes. I clung to it and then clambered aboard by more than human efforts, spurred by danger and sudden love of life. Sitting astride the trough, I paddled the water using my feet as oars and made some progress, though my awkward craft was tossed back and forth by the waves.

The captain had sailed away from the whale quickly, and the ship was vanishing in the distance. I paddled as hard as I could with my feet until night fell. Giving myself up for lost, I still clung to my laundry tub, tossing and dashing about in the wind and the sea. I sailed thus for a day and a night until at last the sea brought me to the coast of a steep island. Vines trailed down its steep cliffs into the water. Exhausted, bruised, and severely bitten by fish, I still managed to climb the vines to the top of the cliff where I fell, dazed, into slumber. . . .

26
Sperm whaling: the capture. (Photo by permission of Forbes Collection, Hart Nautical Museum, MIT)

5

TO FOLLOW THE WHALE: WHALERS & SAILORS

Come all ye young sailormen, listen to me—
I'll sing you a song of the fish of the sea.

Oh, first come the whale, the biggest of all,
He clumb up aloft and let every sail fall.

Then blow ye winds westerly,
westerly blow,
We're bound for the southward,
so steady she goes.

—Boston Come-All-Ye

Farewell to Tarwathie

George Scroggie

This whaling song with a slow, wave-rolling rhythm is a favorite with folksingers today. Whale-lovers will especially like the arrangement sung by Judy Collins, with only the singing of humpback whales for accompaniment. This is the original Scottish version.

Farewell to Tarwathie, adieu Mormond Hill,
Dear land of my fathers, I bid you farewell;
I'm bound for Greenland, and ready to sail
In hopes to find riches in hunting the whale. . . .

27
Whaling picture outside The Whaling Museum, Nantucket, Massachusetts. (Photo courtesy of the Massachusetts Department of Commerce and Development)

Our ship is well rigged and ready to sail,
Our crew they are anxious to follow the whale,
Where the icebergs float and the stormy winds blow,
Where the land and the ocean is covered with snow.

The cold land of Greenland is barren and bare,
No seed-time or harvest is ever known there;
The birds here sing sweetly on mountain and dale,
But there's nae a birdie to sing to the whale.

There's no habitation for man to live there,
The king of that country's the fierce Greenland bear;
There'll be no temptation to tarry long there,
With our ship bumper-full, we'll homeward repair.

— Scottish folklorist Gavin Greig attributes this song to a Scots miller-poet named George Scroggie, who published it in the 1850s.

The Flurry of the Sperm Whale

Frank T. Bullen

The author grew up as a London street urchin and shipped as a cabin boy at the age of twelve. He followed the sea for 15 years more, during which he served as first mate on the New Bedford whaler Cachalot *for a long South Sea voyage. A few years later, Bullen left the sea. Self-educated, he was an avid reader and began writing his recollections of sea-faring in stories and sketches. These stories of "true adventure" were immediately popular, and Bullen was encouraged by other writers, including Kipling.* The Cruise of the Cachalot , *published in 1898, soon became a "boys' adventure classic." It is a straightforward account of everyday life on a whaling voyage, "from a seaman's standpoint." Since in the 1890s Melville had little literary stature or popularity, Bullen had never encountered* Moby Dick *and so thought his book was a first. And in a way, it is, for while it has none of the allegory or mysticism or philosophic depth of Melville's classic, it is an exciting tale of true adventure at sea. . . .*

Through all the vicissitudes of this stange voyage I had hither to felt pretty safe, and as the last thing a man anticipates (if his digestion is all right) is the possibility of coming to grief himself, while fully prepared to see everybody else go under, so I had got to think that whoever got killed I was not to be — a very pleasing sentiment, and one that carries a man far, enabling he him to face dangers with a light heart which otherwise would make a nerveless animal of him.

In this optimistic mood then, I gaily flung myself into my place in the mate's boat one morning, as we were departing in chase of a magnificent cachalot that had been raised just after breakfast. There were no other vessels in sight — much to our satisfaction — the wind was light, with a cloudless sky, and the whale was dead to leeward of us. We sped along at a good rate towards our prospective victim, who was, in his leisurely enjoyment of life, calmly lolling on the surface, occasionally lifting his enormous tail out of water and letting it fall flat upon the surface with a boom audible for miles.

We were, as usual, first boat; but, much to the mate's annoyance, when we were a short half-mile from the whale, our main-sheet parted. It became immediatley necessary to roll the sail up, lest its flapping should alarm the watchful monster, and this delayed us sufficiently to allow the other boats to shoot ahead of us. Thus the second mate got fast some seconds before we arrived on the scene, seeing which we furled sail, unshipped the mast, and went in on him with the oars only. At first the proceedings were quite of the usual character, our chief wielding his lance in most brilliant fashion, while not being fast to the animal allowed us much greater freedom in our evolutions; but that fatal habit of the mate's — of allowing his boat to take care of herself so long as he was setting in some good home-thrusts — once more asserted itself. Although the whale was exceedingly vigorous, churning the sea into yeasty foam over an enormous area, there we wallowed close to him, right in the middle of th turmoil, actually courting disaster.

He had just settled down for a moment, when, glancing over the gunwale, I saw his tail, like a vast shadow, sweeping away from us towards the second mate, who was laying off the other side of him. Before I had time to think, the mighty mass of gristle leapt into the sunshine, curved back from us like a huge bow. Then with a roar it came at us, released from its tension of Heaven knows how many tons. Full on the broadside it struck us, sending every soul but me flying out of the wreckage as if fired from catapults. I did not go because my foot was jammed somehow in the well of the boat, but the wrench nearly pulled my thigh-bone out of its socket. I had hardly released my foot, when, towering above me, came the colossal head of the great creature, as he ploughed through the bundle of debris that had just been a boat. There was an appalling roar of water in my ears, and darkness that might be felt all around. Yet, in the midst of it all, one thought predominated as clearly as if I had been turning it over in my mind in the quiet of my bunk aboard — "What if he should swallow me?" Nor to this day can I understand how I escaped the portals of his gullet, which of course gaped wide as a church door. But the agony of holding my breath soon overpowered every other feeling and thought, till just as something was going to snap inside my head I rose to the surface. I was surrounded by a welter of bloody froth, which make it impossible for me to see; but oh, the air was sweet!

I struck out blindly, instinctively, although I could feel so strong an eddy that voluntary progress was out of the question. My hand touched and clung to a rope, which immediately towed me in some direction — I neither knew nor cared whither. Soon the motion ceased, and, with a seaman's instinct, I grasped, although no definite idea was in my mind as to where it was attached. Presently I came butt up against something solid, the feel of which gathered all my scattered wits into a compact knub of dread. It was the whale! "Any port in a storm," I murmured, beginning to haul away again on my friendly line. By dint of hard work I pulled myself right up the sloping, slippery bank of blubber, until I reached the iron, which, as luck would have it, was planted in that side of the carcass now uppermost. Carcass I said — well, certainly I had no idea of there being any life remaining within the vast mass beneath me; yet I had hardly time to take a couple of turns round myself with the rope (or whale-line, it had proved to be), when I felt the great animal quiver all over, and begin to forge ahead. I was now composed enough to remember that help could not be far away, and that my rescue, providing that I could keep above the water, was but a question of a few minutes. But I was hardly prepared for the whale's next move. Being very near his end, the boat, or

28
The **Cachalot.** *(Drawing from the first edition of* **The Cruise of the Cachalot)**

boats, had drawn off a bit, I supposed, for I could see nothing of them. Then I remembered the flurry. Almost at the same moment it began; and there was I, who with fearful admiration had so often watched the titanic convulsions of a dying cachalot, actually involved in them. The turns were off my body, but I was able to twist a couple of turns round my arms, which, in case of his sounding, I could readily let go.

Then all was lost in roar and rush, as of the heart of some mighty cataract, during which I was sometimes above, sometimes beneath, the water, but always clinging, with every ounce of energy still left, to the line. Now, one thought was uppermost — "What if he should breach?" I had seen them do so when in flurry, leaping full twenty feet in the air. Then I prayed.

Quickly as all the precedings changes had passed came perfect peace. There I lay, still alive, but so weak that, although I could feel the turns slipping off my arms, and knew that I should slide off the slope of the whale's side into the sea if they did, I could make no effort to secure myself. Everything then passed away from me, just as if I had gone to sleep.

I do not at all understand how I kept my position, nor how long, but I awoke to the blessed sound of voices, and saw the second mate's boat alongside. Very gently and tenderly they lifted me into the boat, although I could hardly help screaming with agony when they touched me, so bruised and broken up did I feel. My arms must have been nearly torn from their sockets, for the strands of the whale-line had cut deep into their flesh with the strain upon it, while my thigh was swollen enormously form the blow I received at the onset. Mr. Cruce was the most surprised man I thingk I ever saw. For full ten minutes he stared at me with wide-open eyes. When at last he spoke, it was with difficulty, as if wanting words to express his astonishment. At last he blurted out, "Whar you bin all de time, ennyhaow? 'Cawse ef you bin hangin' on to dat ar wale ev' sence you boat smash, w'y de debbil you hain't all ter bits, hey?" I smiled feebly, but was too weak to talk, and presently went off again into a dead faint.

When I recovered, I was snug in my bunk aboard, but aching in every joint, and as sore as if I had been pounded with a club until I was bruised all over. During the day Mr. Count was kind enough to pay me a visit. With his usual luck, he had escaped without the slightest injury; neither was any other member of the boat's crew the worse for the ducking but myself. He told me that the whale was one of the largest he had ever seen, and as fat as butter. The boat was an entire loss, so completely smashed to pieces that nothing of her or her gear had been recovered. After spending about a quarter of an hour with me, he left me considerably cheered up, promising to look after me in the way of food, and also to send me some books. He told me that I need not worry myself about my inability to be at work, because the old man was not unfavourably disposed towards me, which piece of news gave me a great deal of comfort.

When my poor, weary shipmates came below from their heavy toil of cutting in, they were almost inclined to be envious of my comfort — small blame to them — though I would gladly have taken my place among them again, could I have got rid of my hurts. But I was condemned to lie there for nearly three weeks before I was able to get about once more. In my sleep I would undergo the horrible anticipation of sliding down that awful, cavernous mouth over again, often waking with a shriek, and drenched with sweat.

—Frank T. Bullen,
The Cruise of the Cachalot (1898)

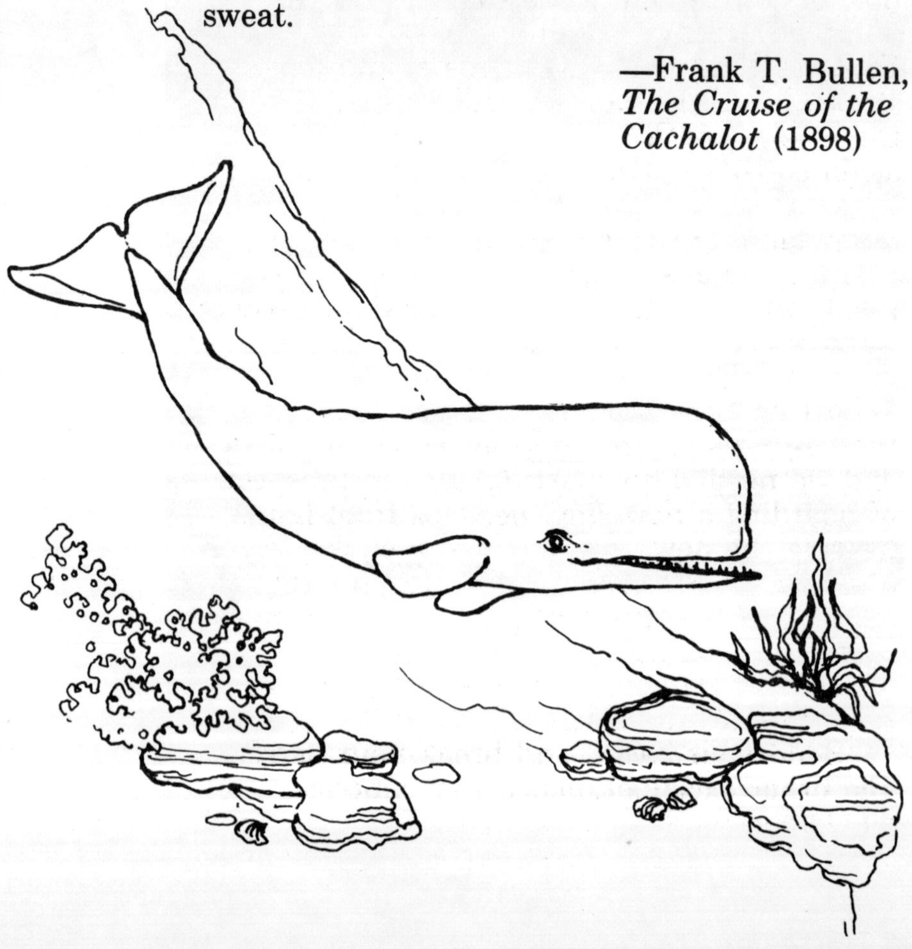

Gamming With a New Bedford Whaler

Richard Henry Dana, Jr.

Tuesday, November 10th, 1835
Santa Barbara, California

Going ashore, as usual, in the gig, just before sundown, to bring off the captain, we found, upon taking in the captain and pulling off again, that our ship, which lay the farthest out, had run up her ensign. This meant "Sail ho!" of course, but as we were within the point we could see nothing. "Give way, boys! Give way! Lay out on your oars, and long stroke!" said the captain; and stretching to the whole length of our arms, bending back again so that our backs touched the thwarts, we sent her through the water like a rocket. A few minutes of such pulling opened the islands, one after another, in range of the point, and gave us a view of the Canal, where was a ship, under top-gallant-sails, standing in, with a light breeze, for the anchorage.

Putting the boat's head in the direction of the ship, the captain told us to lay out again; and we needed no spurring, for the prospect of boarding a new ship, perhaps from home, hearing the news and havingsomething to tell of when we got back, was excitement enough for us, and we gave way with a will. Captain Nye, of the *Loriotte*, who had been an old whaleman, was in the steer-sheets, and fell mightily into the spirit of it.

"Bend your backs, and break your oars!" and other exclamations current among whalemen.

In the meantime it fell flat caim, and, being within a couple of miles of the ship, we expected to board her in few minutes when a breeze sprung up, dead ahead for the ship, and she braced up and stood off toward the islands, sharp on the larboard tack, making good way through the water. This, of course, brought us up, and we had only to "ease larboard oars, pull round starboard!" and go aboard the *Alert*, with something very like a flea in the ear. There was a light landbreeze all night, and the ship did not come to anchor until the next morning.

As soon as her anchor was down we went aboard, and found her to be the whale-ship *Wilmington and Liverpool Packet*, of New Bedford, last from the "off-shore ground," with nineteen hundred barrels of oil. A "spouter" we knew her to be, as soon as we saw her, by her cranes and boats, and by her stump top-gallant masts, and a certain slovenly look to the sails, rigging, spars, and hull; and when we got on board, we found everything to correspond—spouter fashion. She had a false deck, which was rough and oily, and cut up in every direction by the chines of oil casks; her rigging was slack, and turning white, paint worn off the spars and blocks, clumsy seizings, straps without covers, and "homeward-bound splices" in every direction. Her crew, too, were not in much better order. Her captain was a slab-sided Quaker, in a suit of brown, with a broad-brimmed hat, bending his long legs as he

29
Ships **Kutusoff** *and* **Factor** *sperm whaling off Lord Howe's Island. (Watercolor by Benjamin Russell, photo by permission of Forbes Collection, Hart Nautical Museum, MIT)*

moved about decks, with his head down, like a sheep, and the men looked more like fishermen and farmers than they did like sailors.

Though it was by no means cold weather (we having on only our red shirts and duck trousers), they all had on woollen trousers,—not blue and ship-shape, but of all colors,—brown, drab, gray, aye, and *green*,—with suspenders over their shoulders, and pockets to put their hands in. This, added to Guernsey frocks, striped comforters about the neck, thick cowhide boots, woollen caps, and a strong oily smell, and a decidedly green look, will complete the description. Eight or ten were on the fore topsail yard, and as many more in the main, furling the topsails, while eight or ten were hanging about the forecastle, doing nothing. This was a strange sight for a vessel coming to anchor; so we went up to them, to see what was the matter. One of them, a stout, hearty-looking fellow, held out his leg and said he had the scurvy; another had cut his hand; and others had got nearly well, but said that there were plenty aloft to furl the sails, so they were *sogering* on the forecastle. There was only one "splicer" on board, a fine-looking old tar, who was in the bunt of the fore topsail. He was probably the only thorough marline-spike seaman in the ship, before the mast. The mates, of course, and the boat-steerers, and also two or three of the crew, had been to sea before, but only on whaling voyages; and the greater part of the crew were raw hands, just from the bush, and had not yet got the hay-seed out of their hair. The mizzen topsail hung in the buntlines until everything was furled forward. Thus a crew of thirty men were half an hour in doing what would have been done in the *Alert*, with eighteen hands to go aloft, in fifteen or twenty minutes.*

We found they had been at sea six or eight months, and had no news to tell us, so we left them, and promised to get liberty to come on board in the evening for some curiosities. Accordingly, as soon as we were knocked off in the evening and were through supper, we obtained leave, took a boat, and went aboard and spent an hour or two. They gave us pieces of whalebone, and the teeth and other parts of curious sea animals, and we exchanged books with them—a practice very common among ships in foreign ports, by which you get rid of the books you have you have read and re-read, and a supply of new ones in their stead, and Jack is not very nice as to their comparative value. (This visiting between the crews of ships at sea is called, among whalemen, "gamming.")

* (Dana later added a footnote: "I have been told that this description of a whaleman has given offence to the whale-trading people of Nantucket, New Bedford, and the Vineyard. It is not exaggerated; and the appearance of such a ship and crew might well impress a young man trained in the ways of a ship of the style of the *Alert*. Long observation has satisfied me that there are no better seamen, so far as handling a ship is concerned, and none so venturous and skillful navigators, as the masters and officers of our whalemen. But never, either on this voyage, or in a subsequent voyage to the Pacific and

its islands, was it my fortune to fall in with a whaleship whose appearance, and the appearance of whose crew, gave signs of strictness of discipline and seaman-like neatness. Probably these things are impossibilities, from the nature of the business, and I may have made too much of them.")

—Richard Henry Dana, Jr.,
From Two Years Before the Mast *(1840)*

Whale Watchers: Charles M. Scammon

After many years' study of the characteristic habits of the Humpback, we believe that the females of this species resort in large numbers to favorite inland waters, connected with the ocean, to bring forth their young; but there are many exceptions to this rule, incident to their roving disposition. They are found on different coasts in all accessible latitudes, from the new-born calf to the extremely aged. In their wanderings, they are addicted, more than any other rorqual, to "breaching," "bolting," and "finning." In the mating season they are noted for their amorous antics. At such times their caresses are of the most amusing and novel character, and these performances have doubtless given rise to the fabulous tales of the swordfish and thrasher attacking whales. When lying by the side of each other, the megapteras frequently administer alternate blows with their long fins, which love-pats may, on a still day, be heard at a distance of miles. They also rub each other with these same huge and flexible arms, rolling occasionally from side to side, and indulging in other gambols which can easier be imagined than described.

The time of gestation is not known; but in all probability it is the same as that of other large Cetaceans. The calf, when brought forth, is about one-fourth the length of the dam; and it suckles by holding the teat between the extremity of the jaws or lips, while the mother reclines a little on one side; raising the posterior portion of her form nearly out of the water, and lying in a relaxed condition.

From *The Marine Mammals of the North-Western Coast of North America* by Charles Scammon (San Francisco, 1874; reprinted 1968 by Dover Publications, Inc.).

The Reluctant Whaleman

Herman Melville

Most 19th-century sailors found much to complain of—especially the ship's food—but few could describe their situation as well as ersthwhile whaleman Herman Melville.

. . . Our ship had not been many days in the harbour of Nukuheva, in the summer of 1942, before I came to the determination of leaving her. That my reasons for resolving to take this step were numerous and weighty may be inferred from the fact that I chose rather to risk my fortunes among the savages of the island than to endure another voyage on board the *Dolly*. . . .

In numberless instances had not only the implied but the specified conditions of the [ship's] articles been violated on the part of the ship in which I served. The usage on board of her was tyrannical; the sick had been inhumanly neglected; the provisions had been doled out in scanty allowance; and her cruises were unreasonably protracted. The captain was the author of these abuses; it was in vain to think that he would either remedy them, or alter his conduct, which was arbitrary and violent in the extreme. His prompt reply to all complaints and remonstrances was—the butt end of a hand-spike, so convincingly administered as effectually to silence the aggrieved party.

To whom could we apply for redress? We had left both law and equity on the other side of the Cape and unfortunately, with very few exceptions, our crew was composed of a parcel of dastardly and mean-spirited wretches, divided among themselves, and only united in enduring without resistance the unmitigated tyranny of the captain. It would have been mere madness for any two or three of the number, unassisted by the rest, to attempt making a stand against his ill usage. They would only have called down upon themselves the particular vengeance of this "Lord of the Plank," and subjected their shipmates to additional hardships.

But, after all, these things could have been endured awhile, had we entertained the hope of being speedily delivered from them by the due completion of our servitude. But what a dismal prospect awaited us in this quarter! The longevity of Cape Horn whaling voyages is proverbial, frequently extending over a period of four or five years.

Some long-haired, barenecked youths, who, forced by the united influences of a roving spirit and hard times, embark at Nantucket for a pleasure excursion to the Pacific, and whose anxious mothers provide them with bottled milk for the occasion, oftentimes return very respectable middle-aged gentlemen.

The very preparations made for one of these expeditions are enough to fighten one. As the vessel carries out no cargo, her hold is fill with provisions for her own consumption.

The owners, who officiate as caterers for the voyage, supply the larder with an abundance of dainties. Delicate morsels of beef and pork, cut on scientific principles from every part of the animal, and of all conceivable shapes and sizes, are carefully packed in salt, and stored away in barrels; affording a neverending variety in their different degrees of toughness, and and in the peculiarities of their saline properties. Choice old water too, decanted into stout six-barrel casks, and two pints of which is allowed every day to each soul on board; together with ample store of sea-bread, previously reduced to a state of petrification, with a view to preserve it either from decay or consumption in the ordinary mode, are likewise provided for the nourishment and gastronomic enjoyment of the crew.

But not to speak of the quality of these articles of sailors' fare, the abundance in which they are put on board a whaling vessel is almost incredible. Oftentimes, when we had occasion to break out in the hold, and I beheld the successive tiers of casks and barrels, whose contents were all destined to be consumed in due course by the ship's company, my heart has sunk within me.

—*Herman Melville,*
Typee (1846)

30
Children in Nantucket gaze at figurehead from ship. (Photo courtesy of Massachusetts Department of Commerce and Development)

Thar She Blows!
As told by Edmund Fuller

What does a whaleman do when he's home at last? Spins yarns about the whaling life, of course. "Thar She Blows!" is funniest when read aloud, for the old whaleman's growing exasperation is irresistible. . .

It was the good ship Mozambique, Cap'n Symes commandin', his name bein' the same as mine but no kin, thank God. We wuz four months an' twenty days out o' New Bedford an' not a drop o' ile in the tanks. I'm standin' my watch when the man aloft calls out, "Thar she blows!"

"Where away?" sez I.

"Four points off the starboard quarter," sez he.

An' I goes aft.

"Cap'n Symes," I sez, "thar she blows."

"Where away?" sez he.

"Four points off the starboard quarter," sez I. "Shall we lower away?"

Cap'n Symes sez, "It's blown' right too pert. 'Tain't fitten fer to lower. Go for'ard an' stan' yer watch!"

An' I goes for'ard. Purty soon the man aloft calls out, "Thar she blows and spouts!"

An' I goes aft. An' I sez, "Cap'n Symes, thar she blows and spouts. Shall we lower away?"

Thar She Blows!
As told by Edmund Fuller. Taken from *A Treasury of American Folklore* by B.A. Botkin.

31 *This hand-carved weathervane atop a waterfront cupola in Nantucket is a reminder of the days when whalers from the island brought home treasures of sperm oil from their seafaring adventures. (Photo courtesy of the Massachusetts Department of Commerce and Development)*

Cap'n Symes sez, "I tole ye once and I tell ye twice. By the left hind leg o' the Lamb O' God, it's blowin' right too pert, an' 'tain't fitten fer to lower. Go for'ard an' stan' yer watch."

An' I goes for'ard. Purty soon the man aloft calls out, "Thar she blows and spouts, an' breaches an' bellows!"

An' I goes aft. "Cap'n Symes," I sez, "thar she blows an' spouts, an' breaches an' bellows. Shall we lower away?"

Cap'n Symes sez, "I tole ye once and I tole ye twice and I tell ye three times. By the twenty-three legs of the Twelve Apostles, it's blown' right too pert an' 'tain't fitten fer to lower. Go for'ard an' stan' yer watch."

An' I goes for'ard. Purty soon the man aloft calls out, "Thar she blows, an' spouts, an' breaches an' bellows, an' a sparm at that!"

An' I goes aft. "Cap'n Symes," I sez, "thar she blows an' spouts, an' breaches an' bellows, an' a sparm at that. Shall we lower away?"

Cap'n Symes sez, "By the forty-eight fat-cheeked Cherubim that flutter 'round the thrice-sanctified throne of the thrice-sanctified Christ, it's blowin' right too pert, an' 'tain't fitten fer to lower. But if yew want to lower, lower away an' be damned!"

Well, we chased that whale through the first watch an' 'long about eight bells we come alongside and I sez to the men in the boat, "Shall I let 'er have the long dart?" An' they sez, "Sock it to 'er." An' I socked it to 'er an' it tuk! Well sir, when we cut that whale up we took out eighty-six bar'ls o' the finest sparm ile an' forty-eight pounds o' ambergris. That night I'm down in the fo'c'sle when Cap'n Symes comes down the lee gangway an' he sez to me, "Mr. Symes," he sez, "you're the best man out o' New Bedford with the long dart. Down in my cabin there's a box o' fine Hay-vana seegars and a case o' Jee-maica rum. They're yours for the rest o' the v'ige."

"Cap'n Symes, sir," I sez, "I don't want none o' yer fine Hay'vana seegars an' I don't want none o' yer Jee-maica rum. All I want out o' yew, Cap'n Symes sir, is a little Christian civility—an' goddamned little o' that."

As told by Edmund Fuller. Taken from A Treasury of American Folklore *by B.Z. Botkin.*

Letter From an American Farmer: Martha's Vineyard and the Whale Fishery (1782)

J. Hector St. John de Crevecoeur

The French-born author settled in America in about 1765. His observations on America were charming and appreciative.

. . . This island therefore, like Nantucket, is become a great nursery which supplies with pilots and seamen the numerous coasters with which this extended part of America abounds. Go where you will from Nova Scotia to the Mississippi, you will find almost everywhere some natives of these two islands employed in seafaring occupations. Their climate is so favorable to population that marriage is the object of every man's earliest wish; and it is a blessing so easily obtained that great numbers are obliged to quit their native land and go to some other countries in quest of subsistence. The inhabitants are all Presbyterians, which is the established religion of Massachusetts; and here let me remember with gratitude the hospitable treatment I received from B. Norton, Esq., the colonel of the island, as well as from Dr. Mahew, the lineal descendant of the first proprietor. Here are to be found the most expert pilots, either for the great bay, their sound, Nantucket shoals, or the different ports in their neighbourhood. In stormy weather they are always at sea, looking out for vessels, which they board with singular dexterity, and hardly ever fail to bring safe to their intended harbour. Gay-Head, the western point of this island, abounds with a variety of ochres of different colours, with which the inhabitants paint their houses.

The vessels most proper for whale fishing are brigs of about 150 tons burthen, particularly when they are intended for distant latitudes; they always man them with thirteen hands in order that they may row two whale-boats, the crews of which must necessarily consist of six, four at the oars, one standing on the bows with the harpoon, and the other at the helm. It is also necessary that there should be two of these boats, that if one should be destroyed in attacking the whale, the other, which is never engaged at the same time, may be ready to save the hands. Five of the thirteen are always Indians; the last of the complement remains on board to steer the vessel during the action. They have no wages; each draws a certain established share in partnership with the proprietor of the vessel, by which economy they are all proportionably concerned in the success of the enterprise and all equally alert and vigilant.

None of these whalemen ever exceed the age of forty; they look on those who are past that period not to be possessed of all that vigour and agility which so adventurous a

---

Letter From an American Farmer: Martha's Vineyard and the Whale Fishery (1782)
Letter VI of *Letters from an American Farmer,* published 1782.

business requires. Indeed, if you attentively consider the immense disproportion between the object assailed and the assailants, if you think of the diminutive size and weakness of their frail vehicle, if you recollect the treachery of the element on which this scene is transacted, the sudden and unforeseen accidents of winds, etc., you will readily acknowledge that it must require the most consummate exertion of all the strength, agility, and judgement of which the bodies and the minds of men are capable to undertake these adventurous encounters.

As soon as they arrive in those latitudes where they expect to meet with whales, a man is sent up to the masthead; if he sees one, he immediately cries out PAWANA , *here is a whale;* they all remain still and silent until he repeats AWAITE PAWANA , *a whale,* when in less than six minutes the two boats are launched, filled with every implement necessary for the attack. They row toward the whale with astonishing velocity; and as the Indians early became their fellow-labourers in this new warfare, you can easily conceive how the Nattic expressions became familiar on board the whale-boats. Formerly it often happened that whale vessels were manned with none but Indians and the master; recollect also that the Nantucket people understand the Nattic and that there are always five of these people on board,

There are various ways of approaching the whale, according to their peculiar species; and this previous knowledge is of the utmost consequence. When these boats are arrived at a reasonable distance, one of them rests on its oars and stands off as a witness of the approaching engagement; near the bows of the other, the harpooner stands up, and on him principally depends the success of the enterprise. He wears a jacket, closely buttoned, and round his head a handkerchief, tightly bound; in his hands he holds the dreadful weapon, made of the best steel, marked sometimes with the name of their town and sometimes with that of their vessel, to the shaft of which the end of a cord of due strength, coiled up with the utmost care in the middle of the boat, is firmly tied; the other end is fastened to the bottom of the boat. Thus prepared, they row in profound silence, leaving the whole conduct of the enterprise to the harpooner and to the steersman, attentively following their directions. When the former judges himself to be near enough to the whale, that is, at the distance of about fifteen feet, he bids them stop; perhaps she has a calf, whose safety attracts all the attention of the dam, which is a favourable circumstance; perhaps she is of a dangerous species, and it is safest to retire, though their ardour will seldom permit them; perhaps she is asleep — in that case he balances high the harpoon, trying in this important moment to collect all the energy of which he is capable. He launches it forth; she is struck; from her first movement they judge of her temper, as well as of their future success.

Sometimes, in the immediate impulse of rage, she will attack the boat and demolish it with one stroke of her tail; in an instant, the frail vehicle disappears and the assailants are immersed in the dreadful element. Were the whale armed with the jaws of the shark, and as voracious, they never would return home to amuse their listening wives with the interesting tale of the adventure. At other times, she will dive and disappear from human sight; and everything must then give way to her velocity, or else all is lost. Sometimes she will swim away, as if untouched, and draw the cord with such swiftness that it will set the edge of the boat on fire by the friction. If she rises before she has run out the whole length, she is looked upon as a sure prey. The blood she has lost in her flight weakens her so much that if she sinks again, it is but for a short time; the boat follows her course with an almost equal speed. She soon reappears; tired at last with convulsing the element, which she tinges with her blood, she dies and floats on the surface. At other times, it may happen that she is not dangerously wounded, though she carries the harpoon fast in her body, when she will alternately dive and rise, and swim on with unabated vigour. She then soon reaches beyond the length of the cord and carries the boat along with amazing velocity: this sudden impediment sometimes will retard her speed; at other times it only serves to rouse her anger and to accelerate her progress.

The harpooner, with the axe in his hands, stands ready. When he observes that the bows of the boat are greatly pulled down by the diving whale and that it begins to sink deep and to take much water, he brings the axe almost in contact with the cord; he pauses, still flattering himself that she will relax; but the moment grows critical; unavoidable danger approaches; sometimes men more intent on gain than on the preservation of their lives will run great risks; and it is wonderful how far these people have carried their daring courage at this awful moment! But it is vain to hope; their lives must be saved; the cord is cut; the boat rises again. If after thus getting loose she reappears, they will attack and wound her a second time. She soon dies, and when dead, she is towed alongside of their vessel, where she is fastened.

The next operation is to cut with axes and spades every part of her body which yields oil; the kettles are set a-boiling; they fill their barrels as fast as it is made; but as this operation is much slower than that of cutting up, they fill the hold of their ship with those fragments, lest a storm should arise and oblige them to abandon their prize. It is astonishing what a quantity of oil some of these fish will yield and what profit it affords to those who are fortunate enough to overtake them. The river St. Lawrence whale, which is the only one I am well acquainted with, is seventy-five feet long, sixteen deep, twelve in the length of its bone, which commonly weighs 3,000 lb., twenty in the breadth of their tails, and produces 180 barrels of oil: I once saw 16 boiled out of the tongue only. After having once vanquished this leviathan, there are two enemies to be dreaded beside the wind, the first of which is the shark: that fierce, voracious fish, to which nature has given such dreadful offensive weapons, often comes alongside, and in spite

32 *Edgartown, Massachusetts, on the island of Martha's Vineyard, was the prosperous home port of Arctic whaling vessels. During the 19th century, men in the village refined whale oil and made candles, while each year women turned out 15,000 pairs of socks, 3,000 pairs of mittens and 600 wigs. (Photo courtesy of the Massachusetts Department of Commerce and Development)*

of the people's endeavours, will share with them in their prey, at night particularly. They are very mischievous, but the second enemy is much more terrible and irresistible; it is the killer, sometimes called the thrasher, a species of whales about thirty feet long. They are possessed of such a degree of agility and fierceness as often to attack the largest spermaceti whales, and not seldom to rob the fishermen of their prey; nor are there any means of defence against so potent an adversary.

When all their barrels are full, for everything is done at sea, or when their limited time is expired and their stores almost expended, they return home, freighted with their valuable cargo, unless they have put it on board a vessel for the European market. Such are, as briefly as I can relate them, the different branches of the economy practised by these bold navigators and the method with which they go such distances from their island to catch this huge game.

The following are the names and principal characteristics of the various species of whales known to these people:

The river St. Lawrence whale, just described.

The disko, or Greenland ditto.

The right whale, or seven feet bone, common on the coasts of this country, about sixty feet long.

The spermaceti whale, found all over the world and of all sizes; the longest are sixty feet and yield about 100 barrels of oil.

The hump-backs, on the coast of Newfoundland, from forty to seventy feet in length.

The finn-back, an American whale, never killed, as being too swift.

The sulphur-bottom, river St. Lawrence, ninety feet long; they are but seldom killed, as being extremely swift.

The grampus, thirty feet long, never killed, on the same account.

The killer, or thrasher, about thirty feet; they often kill the other whales, with which they are at perpetual war.

The black fish whale, twenty feet; yields from eight to ten barrels.

The porpoise, weighing about 160 lb.

In 1769 they fitted out 125 whalemen; the first fifty that returned brought with them 11,000 barrels of oil. In 1770 they fitted out 135 vessels for the fisheries, at thirteen hands each; four West-Indiamen, twelve hands; twenty-five wood vessels, four hands; eighteen coasters, five hands; fifteen London traders, eleven hands. All these amount to 2,158 hands, employed in 197 vessels. Trace their progressive steps between the possession of a few whale-boats, and that of such a fleet!

The moral conduct, prejudices, and customs of a people who live two thirds of their time at sea, must naturally be very different from those of their neighbours, who live by cultivating the earth. That long abstemiousness to which the former are exposed, the breathing of saline air, the frequent repetitions of danger, the boldness acquired in surmounting them, the very impulse of the winds, to which they are exposed — all these, one would imagine, must lead them, when on shore, to no small desire of inebriation and a more eager pursuit of those pleasures, of which they have been so long deprived and which they must soon forego. There are many appetites that may be gratified on shore, even by the poorest man, but which must remain unsatisfied at sea. Yet, notwithstanding the powerful effects of all these causes, I observe here, at the return of their fleets, no material irregularities, no tumultuous drinking assemblies; whereas in our continental towns, the thoughtless seaman indulges himself in the coarsest pleasures and, vainly thinking that a week of debauchery can compensate for months of abstinence, foolishly lavishes in a few days of intoxication the fruits of half a year's labour. On the contrary, all was peace here, and a general decency prevailed throughout; the reason, I believe, is that almost everybody here is married, for they get wives very young; and the pleasure of returning to their families absorbs every other desire.

The motives that lead them to the sea are very different from those of most other seafaring men; it is neither idleness nor profligacy that sends them to that element; it is a settled plan of life, a well-founded hope of earning a livelihood; it is because their soil is bad, that they are early initiated to this profession; and were they to stay at home, what could they do? The sea therefore becomes to them a kind of patrimony; they go to whaling with as much pleasure and tranquil indifference, with as strong an expectation of success, as a landman undertakes to clear a piece of swamp.

—J. Hector St. John de Crevecoeur,
Letters from an American Farmer (1782)

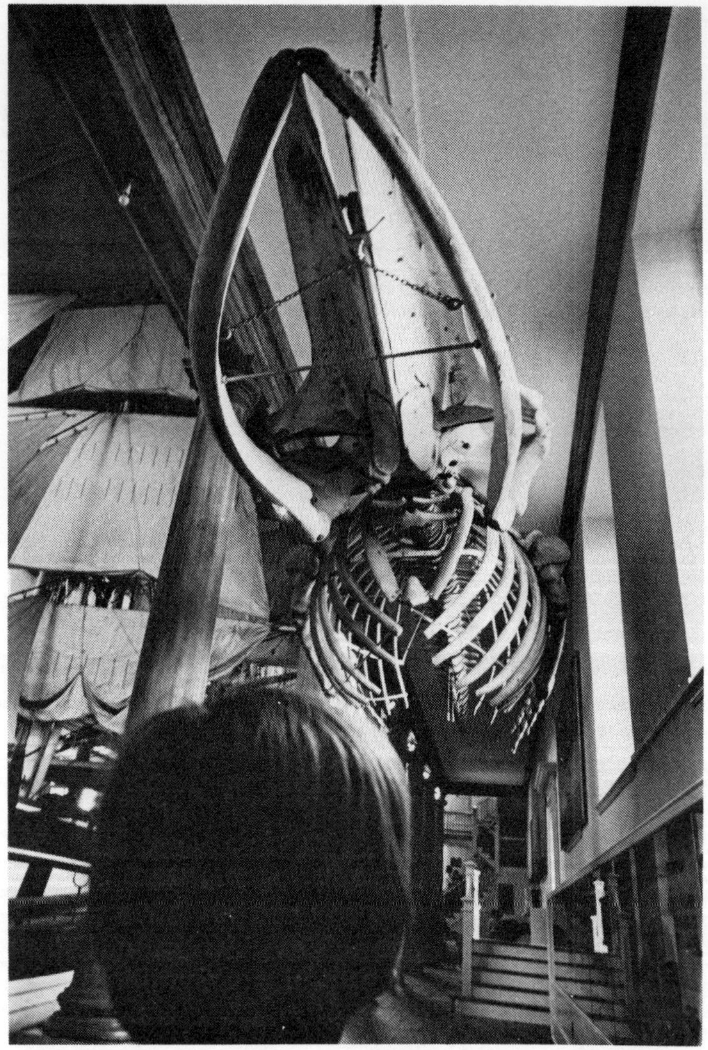

33 *Whale skeleton at the whaling museum on Johnny Cake Hill, New Bedford, Massachusetts. The museum displays relics from the seafaring days of this historic city, which was once the home port of the world's greatest whaling fleet. (Photo courtesy of the Massachusetts Department of Commerce and Development)*

A Whaleman's Dictionary: A Glossary of Whaling Words, Phrases and Slang

Charles M. Scammon

Bailer—A copper or iron vessel used in dipping up oil. Two of these utensils are used on board of a whaler: one with a short, upright handle, called the hand-bailer; the other, with a staff to it six feet long, used at the try-works for bailing the oil from the pots, is called a long-handled bailer.

Becket—A thing used in ships to confine loose ropes, tackles, or spars.

Black-skin—The rete-mucosum and the cuticle, the principal seat of color in whales.

Black-whale, or black-whale oil—Is that produced from all the baleen whales, including the rorquals. All these varieties of whales are sometimes termed black whales, in contradistinction to the Sperm Whale.

Blanket-piece—A strip or section of blubber cut from a whale in a spiral direction, and raised by means of the cutting-tackle. It varies from two to four feet or more in width, and is in length from ten to twenty-five feet.

Blasted—A term used to signify that a whale is much swollen, or far advanced in decomposition after death.

Blink, or ice-blink—stratum of lucid whiteness which appears in the lower part of the atmosphere over ice and land covered with snow.

Blow—Blow signifies the action of the whale in making one respiration.

Blubberfork—A utensil used in pitching the minced blubber from the tubs into the trypots.

Blubber-hook—A stout iron hook of seventy-five to a hundred pounds weight, which is used in flensing a whale.

Boarding-knife—A sharp two-edged instrument, principally used in cutting the toggle-hole in the blubber of a whale, for the purpose of inserting the strap to the mass of fat called the blanket-piece.

Bolting—Signifies the action of a whale when it leaps out diagonally to the surface of the water.

Bone-spade—A cutting-spade, with a thin, long shank to it.

Bonnet—Cheever defines the bonnet of a Right Whale "as being the crest or comb where there burrow legions of barnacles and crabs, like rabbits in a warren, or insets in the shaggy bark of an old tree." (Note: This description applies especially to the southern Right Whales; in the northern Right Whale's bonnet, but very few barnacles are present, and

---

A Whaleman's Dictionary: A Glossary of Whaling Words, Phrases and Slang
From *Marine Mammals of the North-Western Coast of North America,* by Charles M. Scammon (San Francisco, 1874; reprinted 1968 by Dover Publications, Inc.).

comparatively few parasites of any description.)

Bomb-shot—The distance a bomb-lance can be fired into a whale effectively, which is about twenty yards.

Breaching—Signifies the movement of a whale when leaping out of the water, in nearly a perpendicular direction or otherwise.

Breaking black-skin—The act of darting a harpoon into a whale.

Bring-to—The act of a whale when it ceases its progressive motion.

Broken voyage—An unprofitable voyage, or a losing voyage.

Case—The sack or cavity which lies on the right side and upper portion of the head of a Sperm Whale, and which contains oil and spermaceti. These combined constitute what is called "head-matter."

Cooler—A copper or iron tank into which the oil from the try-pots is first bailed.

Cutting-spade—A sharp, flat implement, like a thin chisel, which is fixed to a pole ten or more feet in length, and is used in cutting the blubber from a whale. A "bone-spade" is merely a cutting-spade, with a long, thin shank to it. It is employed in cutting out the throat-bone of a baleen whale. The "headspade" is thicker and heavier than the ordinary cutting-spade, and is used in cutting the skull-bone, which separates the whale's head from the body.

Darting distance —The distance the harpoon is usually thrown effectively by hand, which is about eighteen feet.

Deck-pot—An iron pot of similar shape and size with the try-pots, but having legs to it.

Fast—A term used when the harpoon penetrates the whale, "as the boat is fast," signifying that the boat is fastened to the whale by means of the harpoon attached to the line.

Fins—The pectorals or side fins of a whale are called fins, in contradistinction ot the flukes, or caudal fin.

Finning—The action of a whale when lying partly on its side, raising one fin out of the water, and striking it upon the surface, causing a splash.

Fire-pike—An instrument used in feeding and stirring the fires when trying out oil.

Flukes—The posterior extremities, or caudal fin of a whale.

Flukes, to cut flukes out, or the whale cuts its flukes out—Is a whaler's phrase to describe the action of the animal when it throws its caudal fin sidewise and upward, upon or above the surface of the water, which is an indication that the creature has taken fright, and suddenly endeavors to escape. "He attempted to cut his flukes out," is a slang expression in whaling parlance, when any member of a ship's company become refractory, or attempt in any manner to create disturbance on board.

Gaff, or gaff-hook—A sharp, strong iron hook attached to a short or long wooden handle, and used in handling blubber.

Galley—Frightened; or "The whale is gallied," or "The boar-steerer got so gallied he could not strike the whale." The state of being frightened.

Gam—Signifies two or more ships meeting, and their companies exchanging visits. Gam also signifies a large collection or herd of whalebone whales.

Gamming—To visit from one whaling-vessel to another.

Gurry—Is the term by which the combination of water, oil, and dirt is known when collected upon the ship's deck and below, during the time of cutting in and boiling out a whale.

Head-spade—A thick, heavy cutting-spade used in cutting the head-bones of a whale.

Hopper, or scrap-hopper—A flaring wooden box used as a receptacle for the scraps when taken from the try-pots.

Hose-cock—A large brass cock attached to the end of the oil-hose.

Hose-scuttle—A small opening in a whaling-vessel's deck through which the oil from a cask runs into the hose-tub.

Hose-tub—A large wooden tub, or copper tank, with a strainer in the bottom to which the oil-hose is attached. The tub is lashed close beneath the deck, immediatley under the hose-scuttle.

Hump—A protuberance or elevation upon the top of a whale's body, and generally nearest to the posterior extremity. In some species it appears somewhat like a dorsal fin.

Junk—A wedge-shaped mass of cellular membranous substance, filled with oil and spermaceti, which lies between the case and the white-horse of a Sperm Whale's head.

Knuckle-joint—The joint of the fin, or pectorals, which connects with the shoulderblade of a whale.

Lay on, or lay the boat on—Is to turn the head of the boat toward any object by means of the steering-oar.

Lay off, or lay the boat off—Is to turn the head of the boar from any object, by means of the steering-oar.

Leaning blubber—To cut the flesh, or lean meat, etc., from it.

Leaning-knife—A large knife used in cutting the flesh, or other tissue destitute of oil, from the blubber, preparatory to trying it out.

Lipper—A piece of thin blubber cut into an oblong square, and punctured, so as to admit a man's fingers to hold it, when used in wiping up the gurry from the decks, etc.

Lipper off the decks—Is to wipe the gurry off with lippers.

Lobtailing, or loptailing—The action of a whale when it raises its flukes high out of water, and strikes them down with great force upon the surface. 2d. A whale beating the surface of the water with its caudal fin.

Loose irons—Harpoons which are darted into a whale with no line attached.

Mate—This word is used by whalers to signify a partnership between two vessels in pursuing whales.

Mill—To turn in an opposite direction, or nearly so; as, "The whale was running to

Whaleboat equipment. (Engraving from Scammon's **Marine Mammals of the North-Western Coast)**

windward, but 'milled,' and ran to leeward."

Mincing-knife—A thin, broad steel implement, sharp on one edge, with a handle on each end, which is used in cutting the blubber into thin slices, preparatory to being boiled or tried out.

Mux, or muxing—To perform work in an awkward or improper manner; as, "He made a mux of it, and missed the whale,: i.e., by improper management a whale was not struck, which otherwise might have been secured.

Noddle-end—The anterior and upper portion of a Sperm Whale's head.

Nib-end—The anterior and upper portion of a whalebone whale's head.

Nisket—The anus of a whale.

Oil-hose—A stout hose made of grained leather, for the purpose of conducting the oil from a vessel's deck into empty casks in the hold.

Piece—This word is frequently used for briefness, instead of the word blanketpiece.

Pike—A pointed iron or steel implement fixed to the end of a pole, and used in pitching or handling the pieces of blubber, etc.

Raising whales—When first seeing a whale from the mast-head, or other place, it is termed "raising a whale." It is also used to express the fact of seeing any object, as, "We raised the land, sixty miles distant."

Recruits—Fresh provisions and supplies for a whale-ship. 2d. Various articles put on board the ship to purchase recruits with (at remote islands or coasts), such as clothing, calicoes, tobacco, and a variety of other domestic articles.

Ridge—The upper portion of a whale's "small."

Ripple—The agitation of the surface of the water by a whale when swimming near it, but still not visible, as, "I can not see the whale, but can see its ripple."

Rounding, or rounding out—The attitude of a whale when curving its small in order to descend below the surface of the water.

Rugged—The state of the sea when agitated. A rough sea, accompanied with blowing weather, is termed by whalers "rugged weather."

Sampson-post—A heavy upright timber, firmly secured in the deck (and extending about two feet above it), to which the fluke-chain, or fluke-rope, is made fast.

Scooping—This term is applied to the action of whalebone when feeding.

Scrap-hopper—A flaring wooden box which is placed beside the try-works to receive the scraps when skimmed from the try-pots. It has holes in the bottom through which the oil that drips from the scraps runs into a tub called a "strainer-cooler."

Scraps—The residue of the blubber after the oil is extracted by boiling.

Scrimshawing, or to scrimshaw, or skimshander, as sometimes termed—Is to execute any piece of ingenious mechanical work. It is applied particularly to polishing and engraving upon whalebone or whale's teeth, or manufacturing fancy articles from the same materials.

Skimmer—A broad scoop composed of copper, iron, and wood. The utensil is about seven feet long and is used in removing scraps from the try-pots.

Skimming-slicks—A phrase used by whalemen when employed in saving any small amount of fat from the carcass after the blubber has been stripped off.

Slumgullion—The offal from the blubber of a whale.

Small—The posterior portion of a whale, between the vent and caudal fin.

Sounding—Is the term signifying the descent of the whale after being fastened to.

Stern, or stern all—Is the invariable order given to the whaleboat's crew when it becomes necessary to propel the boat stern foremost.

Stirring-pole—A wooden pole six or seven feet long, used to stir the blubber in the try-pots.

Stopped—To confine a rope, or ropes, usually by a smaller line, as, "Stop the line to the 'iron-pole'" (or harpoon staff), i.e.,

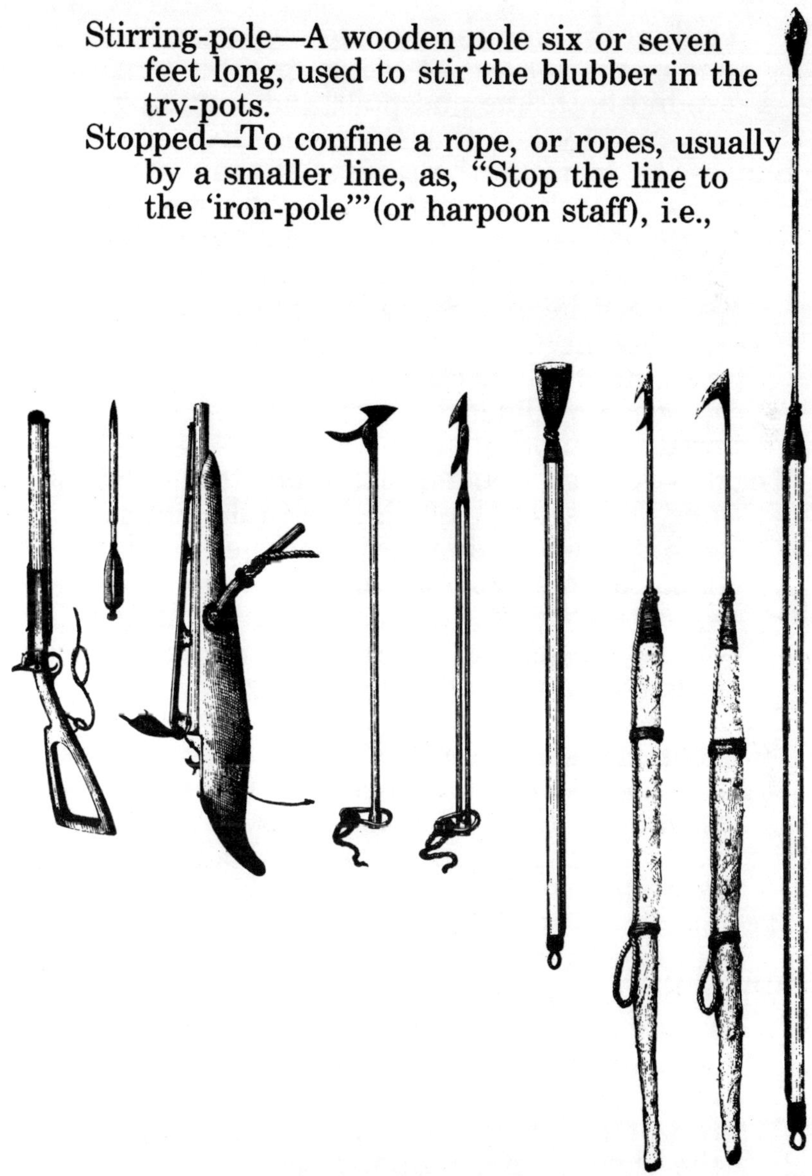

34 *Implements belonging to a whale boat. From left to right: hand-lance, bomb-lance gun, bomb-lance, Greener's harpoon-gun, two versions of Greener's-gun-harpoon, boat-spade, toggle-harpoon, one-flued harpoon. (Engraving from Scammon's* **Marine Mammals of the North-Western Coast**)

confine the line to the pole by passing one or more turns of twine, or rope-yarn, around both line and pole, and confining the ends by knotting them together.

Suds—The foam on the surface of the water caused by the violent actions of a whale in the course of its capture, as, "In order to get a good chance to kill the whale, you must keep close to the suds."

Sweeping—The action of a whale when wielding its flukes from side to side in an offensive or defensive manner, causing a great commotion on the water.

Toggle—A large wooden pin, about three feet long, and six or eight inches in diameter, with a swell or shoulder near the middle. It has a hole near one end, through which a rope is attached, which is termed the toggle-lanyard. This lanyard is used in handling or confining the toggle.

Turning flukes—The last action of a whale upon the surface of the water when making its descent beneath, which is to elevate its caudal fin in the air as the body assumes nearly a perpendicular attitude.

Trying out—To extract the oil from the blubber by boiling.

White-horse—An extremely tough and sinewy substance resembling blubber, but destitute of oil, which lies between the upper jaw and junk of a Sperm Whale.

White-water—The foam or spray causd by the violent actions of a whale.

—Charles M. Scammon, *Marine Mammals of the North-Western Coast of North America* (1874)

35
Half-scale replica of the whaler **Lagoda,** *at the New Bedford Whaling Museum. (Photo courtesy of the Massachusetts Department of Commerce and Development)*

The Towns That Whaling Built: Nantucket and New Bedford

A large part of the romance of whaling centres around the island of Nantucket and its hardy seamen. It was from here that the Red Men first sallied out in canoes to chase the whale; from here the small sloops first set out laden with cobblestones, as the story goes, to throw at the whales to see if they were near enough to risk a harpoon. These daring Nantucketers were, in 1791, the first to sail to the Pacific, and later on in 1820 to the coast of Japan, and finally they made their ships known in every harbour of the world. Thirty islands and reefs in the Pacific are named after Nantucket captain and merchants.

There is an amusing legend concerning the origin of the island. A giant was said to be in the habit of sleeping on Cape Cod, because its peculiar shape fitted him when he curled himself up. One night he became very restless and thrashed his feet around so much that he got his moccasins filled with sand. In the morning he took off first one moccasin and them the other, flinging their contents across the sea, thus forming th islands of Martha's Vineyard and Nantucket.

From the time of the settlement of the island, the entire population, from the oldest inhabitant down to the youngest child, realized that on the whaling industry depended their livelihood. A story is told of a Nantucket youngster who tied his mother's darning cotton to a fork, and, hurling it at the cat as she tried to escape, yelled "Pay out, mother! Pay out! There she 'sounds' through the window!" The inhabitants always alluded to a train as "tying up," a wagon was called a "side-wheeler," every one you met was addressed as "captain," and a horse was always "tackled" instead of harnessed. The refrain of an old Nantucket song runs as follows:—

> "So be cheery, my lads, let your
> hearts never fail,
> While the bold harpooner is striking
> the whale!"

A young man who had not doubled the cape or harpooned a whale had no chance of winning a Nantucket, New Bedford, or New London belle, and it is stated as a fact that the girls of Nantucket at one time formed a secret society, and one of their pledges was never to marry a man until he had "struck his whale."

. . . Early history tells us that Thomas Macy purchased the island for thirty pounds and two beaver hats, "One for myself and one for my wife," and to him therefore belongs the honor of the settlement of Nantucket; he had been driven away from Massachusetts for

---

The Towns that Whaling Built: Nantucket and New Bedford From *Whale Fishery of New England,* published 1915 by State Street Trust Company, Boston. Reprinted 1968 by Reynolds-DeWalt Printing, Inc., New Bedford, Mass.

sheltering Quakers, which was at that time against the law, and with his friend Edward Starbuck fled to the island and established a colony composed of such well-known families as the Coffins, Husseys, Swaynes, Gardners, Chases, Folgers, and Starbucks. These men were not whalers, but they watched the Indians and learned much from them, and later on employed Ichabod Paddock to come over from Cape Cod and instruct them further.

The character of the island and its situation far out in the ocean, its poor soil, and the number of whales along its shores, all proved an inducement to the Nantucketers to follow the sea as a calling. At first, there were so many whales that they did not find it necessary to go beyond the coast; so, under the guidance of Paddock, lookouts were erected along the South shore, and each man patrolled a certain amount of territory. Each one took his share of whales killed, and business flourished. This method of whaling continued until 1712, when Christopher Hussey, while cruising along the coast, was blown out to sea. He ran across a sperm whale, which he finally killed and brought home. This year was epoch making, as this was the first sperm whale known to have been taken by Americans. The oil from this species of whale being superior to that of all others, the Nantucketers now (1715) decided to change their methods and to whale in the "deep." As the vessels steadily increased in size with greater and greater cargo-carrying capacity, voyages necessarily became longer, extending even to periods of four or five years. In fact, a voyage lasting but two years was considered unusually short. The point of view of most whalers regarding a two-year voyage is shown by the captain who, when boarding his ship, was reminded by a friend that he had not said "Good-by" to his wife,—

"Why should I?" said he; "I am only to be gone two years."

About 1730 "try-works" were built on the vessels instead of on the shore, and the oil was boiled and stowed away at sea, thus allowing the ships to make much longer voyages. At this time Nantucket owned as many whaleships as all the other ports of America combined. Whaling continued to increase, and the sterile island was turned into a prosperous community, when the Revolution came on, and for the time being practically put an end to the industry....

The redoubtable Nantucketers resumed their whaling at the close of the Revolution, and their energy and skill were again yielding rich profits when the War of 1812 almost annihilated the island's fleet. But as it was another case of whale or starve, Nantucket continued to send out a few whalers, and was the only American port during the war that dared to brave the risks of British capture.

About this time, in one of the Pacific ports, an incident occurred which showed in an amusing light the ready wit and intrepid courage of an American whaleman. He had in some way displeased an English naval officer, who, feeling himself highly insulted, promptly challenged the Yankee, who accepted and, being the challenged party, had the choice of weapons. He selected, of course, the weapon with which he was most skilful and took his stand with a poised harpoon. It had altogether too dangerous an appearance for the irate Englishman, particularly as the whaleman was evidently an expert in the manual of thrust and parry, and so with as good grace as he could command, the Englishman withdrew from the fight.

At a very early day in the fishery, whaling vessels, which were at first long rowboats and later small sloops, began to increase in size, and about 1820 ships of three hundred tons were found profitable. The increase in profit producing capacity, strange as it may appear, actually sounded the death-knell of the Nantucket whaling, for across the mouth of the harbour ran a bar, over which it soon became impossible for whaling vessels of large size to pass. The difficulty was for a time overcome by the true Yankee ingenuity of some inventive Nantucketer, who devised the "camel," a veritable dry-dock barge in which the larger whaleships, lightened often of oil and bone, were floated over the bar into the forest of masts. . .

As whaleships still continued to increase in size, the "camel" expedient was only a temporary success; for the time came when vessels were of too great tonnage to be thus floated over the bar, and the daring and skilful Nantucketer, who had taught the civilized world not only how, but where, to whale, had to admit defeat and gradually give up the industry to more fortunately situated ports. At this time, about 1830, Nantucket was commercially the third largest city in Massachusetts, Boston being first and Salem second.

In 1843 Nantucket owned its record number of ships, eighty-eight. In 1846, which is referred to as the "boom" year in American whaling, sixteen vessels cleared from Nantucket and sixty-nine from her near-by rival—New Bedford. In 1869 Nantucket sent her last ship and disappeared from the list of whaling ports. The great fire of 1846 also contributed to the downfall of the industry.

A new era in whaling was to be born, with New Bedford as the centre, and Nantucket was to become only a health resort and mecca for sight-seers, more than ten thousand persons visiting the island in 1914.

New Bedford. New Bedford undoubtedly owed its whaling success to is proximity to Nantucket, to its wonderful harbour, and to the honesty, thrift, and good business ability of its citizens, most of whom were Quakers.

As in Nantucket, the whole city lived to go whaling, and as each inhabitant made more money, he moved his residence higher up on the Hill. It is said that there was an inn called the "Crossed Harpoons," and another callled "Spouter Inn," and there is a Whaleman's Chapel on Johnny Cake Hill, where regular Sunday services were held, at which the following hymn was always sung by the congregation:—

"The ribs and terrors of the whale
  Arched over me in dismal gloom,
While all God's sun-lit waves rolled by
  And left me deepening down to doom.

"I saw the opening maw of hell,
With endless pains and sorrows there;
Which none but they that feel can tell—
Oh, I was plunging to despair—

"In black distress I called to God,
When I could scarce believe him mine, He bowed his ear to my complaints— No more the whale did me confine."

The pulpit of this chapel was made to represent the prow of a whaleship, and was ascended by means of a rope ladder, which the minister, who had been a harpooner in his youth, hauled up after him. *Around the walls of this little church can still be seen tablets erected in memory of many whalemen who lost their lives at sea. There also was a daily paper called *The Whaleman*, which gave the reports of the whaleships and the whaling news. It has been said that New Bedford fathers gave whales for dowers to their daughters, and that they had reservoirs of oil in their attics to burn on gala occasions . . . .

Whaling was a tremendous financial gamble, and until a vessel came home "clean" or "greasy," meaning empty or full, the success of the voyage was not known. They tell a story of a New Bedford captain who had been out for nearly four years, and as he came up to the wharf the owners asked him what luck he had had. His reply was, "I didn't get any whales, but I had a damn good sail."

36
Described in Moby Dick, the Seamen's Bethel, in New Bedford's historic waterfront district, has a pulpit shaped like the prow of a ship. Marble memorial tablets pay tribute to the "men who chased whales in faraway waters." (Photo courtesy of the Massachusetts Department of Commerce and Development)

The real founder of New Bedford, and the pioneer of the whale fishery at this port, was Joseph Russell, who sent his ships out in 1765. Several years later the first ship was launched and was called the "Dartmouth," and this vessel is well known to history owing to the fact that she was one of the ships that carried into Boston Harbour the tea that was thrown overboard. The whaling industry increased steadily, except during the wars, until 1857, when the New Bedford fleet numbered three hundred and twenty-nine vessels, was valued at over twelve million dollars, and employed over twelve thousand seamen. If these vessels had been strung out in line, they would have stretched over ten miles. In addition to these sailors, thousands of others were employed at home making casks, irons, ropes, and many other articles used in whaling. In fact, it was often stated that the population was divided into three parts,—those away on a voyage, those returning, and those getting ready for the next trip.

There were many nationalities represented in the crews of the whalers, and the New Bedford streets presented a very foreign appearance, with Spaniards, Portuguese, Dutch, Norwegians, Germans, French, English, Scotch, Irish, Sandwich Islanders, and New Englanders at every turn. A large number of Portuguese served on whaleships, and a part of the city near the south end of Water Street became known as Fayal.

The "Golden Age," as it is called, of whaling was between 1825 and 1860, and during the whole of this period New Bedford assumed the lead, even long after other ports had given up the pursuit. It is estimated that about the year 1848 there were over seventy millions invested in the industry and seventy thousand persons derived from it their subsistence.

It is an interesting fact that the insurance on American whalemen was about one-half the rate that was charged the Englishman, which certainly showed the superiority of our Yankee seamen. There were several whaling insurance companies in New Bedford. There is a story told of a New Bedford ship owner who had just heard that his vessel had gone down and he hadn't yet received the insurance policy from the company. He sent a letter down to the office which read as follows: "I have heard from my ship and thee need not place the insurance." Of course, the policy was sent up immediately.

The New Bedford whalers explored new grounds, and to this fact chiefly is due the continued prosperity of its whale fishery, but it was to die slowly; in 1875 the fleet from this port had declined to 116 vessels, in 1886 to 77 ships, and in 1906 to 24.

—*Whale Fishery of New England* (1915)

37
Sperm whaling on the bark **President.** *(Watercolor by Benjamin Russell, (Photo by permission of Forbes Collection, Hart Nautical Museum, MIT)*

The Alarmed Skipper

James Thomas Fields

Many a long, long year ago,
Nantucket skippers had a plan
Of finding out, though "lying low,"
How near New York their schooners ran.

They greased the lead before it fell,
And then, by sounding through the night,
Knowing the soil that stuck, so well,
They always guessed their reckoning right.

A skipper gray, whose eyes were dim,
Could tell, by *tasting*, just the spot,
And so below he'd "dowse the glim" —
After, of course, his "something hot."

Snug in his berth, at eight o' clock,
This ancient skipper might be found;
No matter how his craft would rock,
He slept—for skippers' naps are sound!

The watch on deck would now and then
Run down and wake him, with the lead;
He'd up, and taste, and tell the men
How many miles they went ahead.

One night, 'twas Jonathan Marden's watch,
A curious wag—the peddler's son—
And so he mused (the wanton wretch)
"Tonight I'll have a grain of fun.

"We're all a set of stupid fools
To think the skipper knows by *tasting*
What ground he's on—Nantucket schools
Don't teach such stuff, with all their basting!"

And so he took the well-greased lead
And rubbed it o'er a box of earth
That stood on deck—a parsnip bed—
And then he sought the skipper's berth.

"Where are we now, sir? Please to taste."
The skipper yawned, put out his tongue,
Then ope'd his eyes in wondrous haste,
And then upon the floor he sprung!

The skipper stormed and tore his hair,
Thrust on his boots, and roared to Marden,
"Nantucket's sunk; and here we are
Right over old Marm Hackett's garden!"

—*James Thomas Fields*

The Chase: First Day

Herman Melville

Moby-Dick, that classic whale, taunts the driven Captain Ahab into the final, destructive pursuit. . . .

That night, in the mid-watch, when the old man — as his wont at intervals — stepped forth from the scuttle in which he leaned, and went to his pivot-hole, he suddenly thrust out his face fiercely, snuffing up the sea air as a sagacious ship's dog will, in drawing nigh to some barbarous isle. He declared that a whale must be near. Soon that peculiar odor, sometimes to a great distance given forth by the living Sperm Whale, was palpable to all the watch; nor was any mariner surprised when, after inspecting the compass, and then the dog-vane, and then ascertaining the precise bearing of the odor as nearly as possible, Ahab rapidly ordered the ship's course to be slightly altered, and the sail to be shortened.

The acute policy dictating these movements was sufficiently vindicated at daybreak, by the sight of a long sleek on the sea directly and lengthwise ahead, smooth as oil, and resembling in the pleated watery wrinkles bordering it, the polished metallic-like marks of some swift tide-rip, at the mouth of a deep, rapid stream.

"Man the mast-heads!"

Thundering with the butts of three clubbed handspikes on the forecastle deck, Daggoo roused the sleepers with such judgement claps that they seemed to exhale from the scuttle, so instantaneously did they appear with their clothes in their hands.

"What d'ye see?" cried Ahab, flattening his face to the sky.

"Nothing, nothing, Sir!" was the sound hailing down in reply.

"T'gallant sails! — stunsails! alow and aloft, and on both sides!"

All sail being set, he now cast loose the life-line, reserved for swaying him to the main royal-mast head; and in a few moments they were hoisting him thither, when, while two thirds of the way aloft, and while peering ahead through the horizontal vacancy between the main-top-sail and top-gallant-sail, he raised a gull-like cry in the air, "There she blows! — there she blows! A hump like a snow-hill! It is Moby Dick!"

Fired by the cry which seemed simultaneously taken up by the three look-outs, the men on deck rushed to the rigging to behold the famous whale they had so long been pursuing. Ahab had now gained his final perch, some feet above the other look-outs, Tashtego standing just beneath him on the cap of the top-gallant-mast, so that the Indian's head was almost on a level with Ahab's heel. From this height the whale was now seen some mile or so ahead, at every roll of the sea revealing his high sparkling hump,

---

The Chase: First Day
From *Moby Dick, or the White Whale* (1851).

and regularly jetting his silent spout into the air. To the credulous mariners it seemed the same silent spout they had so long ago beheld in the moonlit Atlantic and Indian Oceans.

"And did none of ye see it before?" cried Ahab, hailing the perched men all around him.

"I saw him almost that same instant, Sir, that Captain Ahab did, and I cried out," said Tashtego.

"Not the same instant; not the same — no, the doubloon is mine, Fate reserved the doubloon for me. *I only! none of ye could have raised the White Whale first.* There she blows! there she blows!—there she blows! There again—there again!" he cried, in long-drawn, lingering, methodic tones, attuned to the gradual prolongings of the whale's visible jets. "He's going to sound! In stunsails! Down top-gallant-sails! Stand by three boats. Mr. Starbuck, remember, stay on board, and keep the ship. Helm there! No, no; only black water! All ready the boats there? Stand by, stand by! Lower me, Mr. Starbuck; lower, lower—quick, quicker!" and he slid through the air to the deck.

"He is heading straight to leeward, Sir," cried Stubb, "right away from us; cannot have seen the ship yet."

"Be dumb, man! Stand by the braces! Hard down the helm!—brace up! Shiver her! — shiver her! So; well that! Boats, boats!"

Soon all the boats but Starbuck's were dropped; all the boat-sails set—all the paddles plying; with rippling swiftness, shooting to leeward; and Ahab heading the onset. A pale, death-glimmer lit up Fedallah's sunken eyes; a hideous motion gnawed his mouth.

Like noiseless nautilus shells, their light prows sped through the sea; but only slowly they neared the foe. As they neared him, the ocean grew still more smooth; seemed drawing a carpet over its waves; seemed a noon-meadow, so serenely it spread. At length the breathless hunter came so nigh his seemingly unsuspecting prey, that his entire dazzling hump was distinctly visible, sliding along the sea as if an isolated thing, and continually set in a revolving ring of finest, fleecy, greenish foam. He saw the vast, involved wrinkles of the slightly projecting head beyond. Before it, far out on the soft Turkish-rugged waters, went the glistening white shadow from his broad, milky forehead, a musical rippling playfully accompanying the shade; and behind, the blue waters interchangeably flowed over into the moving valley of his steady wake; and on either hand bright bubbles arose and danced by his side. But these were broken again by the light toes of hundreds of gay fowl softly feathering the sea, alternate with their fitful flight; and like to some flagstaff rising from the painted hull of an argosy, the tall but shattered pole of a recent lance projected from the White Whale's back; and at intervals one of the cloud of soft-toed fowls hovering, and to and fro skimming like a canopy over the fish, silently perched and rocked on this pole, the long tail feathers streaming like pennons.

A gentle joyousness — a mighty mildness of repose in swiftness, invested the gliding whale. Not the white bull Jupiter swimming away with ravished Europa clinging to his graceful horns; his lovely, leering eyes sideways intent upon the maid; with smooth bewitching fleetness, rippling straight for the nuptial bower in Crete; not Jove, not that great majesty Supreme! did surpass the glorified White Whale as he so divinely swam.

On each soft side — coincident with the parted swell, that but once leaving him, then flowed so wide away — on each bright side, the whale shed off enticings. No wonder there had been some among the hunters who namelessly transported and allured by all this serenity, had ventured to assail it; but had fatally found that quietude but the vesture of tornadoes. Yet calm, enticing calm, oh, whale! thou glidest on, to all who for the first time eye thee, no matter how many in that same way thou may'st have bejuggled and destroyed before.

And thus, through the serene tranquillities of the tropical sea, among waves whose hand-clappings were suspended by exceeding rapture, Moby Dick moved on, still withholding from sight the full terrors of his submerged trunk, entirely hiding the wrenched hideousness of his jaw. But soon the fore part of him slowly rose from the water; for an instant his whole marbleized body formed a high arch, like Virginia's Natural Bridge, and warningly waving his bannered flukes in the air, the grand god revealed himself, sounded, and went out of sight. Hoveringly halting, and dipping on the wing, the white sea-fowls longingly lingered over the agitated pool that he left.

With oars apeak, and paddles down, the sheets of their sails adrift, the three boats now stilly floated, awaiting Moby Dick's reappearance.

"An hour," said Ahab, standing rooted in his boat's stern; and he gazed beyond the whale's place, towards the dim blue spaces and wide wooing vacancies to leeward. It was only an instant; for again his eyes seemed whirling round in his head as he swept the watery circle. The breeze now freshened; the sea began to swell.

"The birds! — the birds!" cried Tashtego.

In long Indian file, as when herons take wing, the white birds were now all flying towards Ahab's boat; and when within a few yards began fluttering over the water there wheeling round and round, with joyous, expectant cries. Their vision was keener than man's; Ahab could discover no sign in the sea. But suddenly as he peered down and down into its depths, he profoundly saw a white living spot no bigger than a white weasel, with wonderful celerity uprising, and magnifying as it rose, till it turned, and then there were plainly revealed two long crooked rows of white, glistening teeth, floating up from the undiscoverable bottom. It was Moby Dick's mouth and scrolled jaw; his vast, shadowed bulk still half blending with the blue of the sea. The glittering mouth yawned beneath the boat like an open-doored marble tomb; and giving one sidelong sweep with his steering oar, Ahab whirled the craft aside from this tremendous apparition. Then, calling upon Fedallah to change places with him, went forward to the bows, and seizing Perth's harpoon, commanded his crew to grasp their oars and stand by to stern.

37 An early glass-plate photograph of a whaleboat and its crew. (Photo courtesy of Mystic Seaport, Inc.)

Now, by reason of this timely spinning round the boat upon its axis, its bow, by anticipation, was made to face the whale's head while yet under water. But as if perceiving this stratagem, Moby Dick, with that malicious intelligence ascribed to him, sidelingly transplanted himself, as it were, in an instant, shooting his pleated head lengthwise beneath the boat.

Through and through; through every plank and each rib, it thrilled for an instant, the whale obliquely lying on his back, in the manner of a biting shark, slowly and feelingly taking its bows full within his mouth, so that the long, narrow, scrolled lower jaw curled high up into the open air, and one of the teeth caught in a row-lock. The bluish pearl-white of the inside of the jaw was within six inches of Ahab's head, and reached higher than that. In this attitude the White Whale now shook the slight cedar as a mildly cruel cat her mouse. With unastonished eyes Fedallah gazed, and crossed his arms; but the tiger-yellow crew were tumbling over each other's heads to gain the uttermost stern.

And now, while both elastic gunwales were springing in and out, as the whale dallied with the doomed craft in this devilish way; and from his body being submerged beneath the boat, he could not be darted at from the bows, for the bows were almost inside of him, as it were; and while the other boats involuntarily paused, as before a quick crisis impossible to withstand, then it was that monomaniac Ahab, furious with this tantalizing vicinity of his foe, which placed him all alive and helpless in the very jaws he hated; frenzied with all this, he seized the long bone with his naked hands, and wildly strove to wrench it from its gripe. As now he thus vainly strove, the jaw slipped from him; the frail gunwales bent in, collapsed, and snapped, as both jaws, like an enormous shears, sliding further aft, bit the craft completely in twain, and locked themselves fast again in the sea, midway between the two floating wrecks. These floated aside, the broken ends drooping, the crew at the stern-wreck clinging to the gunwales and striving to hold fast to the oars to lash them across.

At that preluding moment, ere the boat was yet snapped, Ahab, the first to perceive the whale's intent, by the crafty upraising of his head, a movement that loosed his hold for the time; at that moment his hand had made one final effort to push the boat out of the bite. But only slipping further into the whale's mouth, and tilting over sideways as it slipped, the boat had shaken off his hold on the jaw; spilled him out of it, as he leaned to the push; and so he fell flat-faced upon the sea.

Ripplingly withdrawing from his prey, Moby Dick now lay at a little distance, vertically thrusting his oblong white head up and down in the billows; and at the same time slowly revolving his whole spindled body; so that when his vast wrinkled forehead rose — some twenty or more feet out of the water — the now rising swells, with all their confluent waves, dazzlingly broke against it; vindictively tossing their shivered spray still higher into the air. * So, in a gale, the but

* This motion is peculiar to the Sperm Whale.

half baffled Channel billows recoil from the base of the Eddystone, only triumphantly to overleap its summit with their scud.

But soon resuming his horizontal attitude, Moby Dick swam swiftly round and round the wrecked crew; sideways churning the water in his vengeful wake, as if lashing himself up to still another and more deadly assault. The sight of the splintered boat seemed to madden him, as the blood of grapes and mulberries cast before Antiochus's elephants in the book of Maccabees. Meanwhile, Ahab half smothered in the foam of the whale's insolent tail, and too much of a cripple to swim, — though he could still keep afloat, even in the heart of such a whirlpool as that; helpless Ahab's head was seen, like a tossed bubble which the least chance shock might burst. From the boat's fragmentary stern, Fedallah incuriously and mildly eyed him; the clinging crew, at the other drifting end, could not succor him; more than enough was it for them to look to themselves. For so revolvingly appalling was the White Whale's aspect, and so planetarily swift the ever-contracting circles he made, that he seemed horizontally swooping upon them. And though the other boats, unharmed, still hovered hard by; still they dared not pull into the eddy to strike, lest that should be the signal for the instant destruction of the jeopardized castaways, Ahab and all; nor in that case could they themselves hope to escape. With straining eyes, then, they remained on the outer edge of the direful zone, whose centre had now become the old man's head.

Meantime from the beginning all this had been descried from the ship's mast-heads; and squaring her yards, she had borne down upon the scene; and was now so nigh, that Ahab in the water hailed her;—"Sail on the"—but that moment a breaking sea dashed on him from Moby Dick, and whelmed him for the time. But struggling out of it again, and chancing to rise on a towering crest, he shouted,—"Sail on the whale!—Drive him off!"

The Pequod's prows were pointed; and breaking up the charmed circle, she effectually parted the White Whale from his victim. As he sullenly swam off, the boats flew to the rescue.

Dragged into Stubb's boat with blood-shot, blinded eyes, the white brine caking in his wrinkles; the long tension of Ahab's bodily strength did crack, and helplessly he yielded to his body's doom: for a time, lying all crushed in the bottom of Stubb's boat, like one trodden under foot of herds of elephants. Far inland, nameless wails came from him, as desolate sounds from out ravines.

But this intensity of his physical prostration did but so much the more abbreviate it. In an instant's compass, great hearts sometimes condense to one deep pang, the sum total of those shallow pains kindly diffused through feebler men's whole lives. And so, such hearts, though summary in each

---

It receives its designation (pitchpoling) from its being likened to that preliminary up-and-down poise of the whale-lance, in the exercise called pitchpoling, previously described. By objects may be encircling him.

one suffering; still, if the gods decree it, in their life-time aggregate a whole age of woe, wholly made up of instantaneous intensities; for even in their pointless centres, those noble natures contain the entire circumferences of inferior souls.

"The harpoon," said Ahab, half way rising, and draggingly leaning on one bended arm—"is it safe?"

"Aye, Sir, for it was not darted; this is it," said Stubb, showing it.

"Lay it before me;—any missing men?"

"One, two, three, four, five;—there were five oars, Sir, and here are five men."

"That's good.—Help me, man; I wish to stand. So, so, I see him! there! there! going to leeward still; what a leaping spout!—Hands off from me! The eternal sap runs up in Ahab's bones again! Set the sail; out oars; the helm!"

It is often the case that when a boat is stove, its crew, being picked up by another boat, help to work that second boat; and the chase is thus continued with what is called double banked oars. It was thus now. But the added power of the boat did not equal the added power of the whale, for he seemed to have treble-banked his every fin; swimming with a velocity which plainly showed, that if now, under these circumstances, pushed on, the chase would prove an indefinitely prolonged, if not a hopeless one; nor could any crew endure for so long a period, such an unintermitted, intense straining at the oar; a thing barely tolerable only in some one brief vicissitude. The ship itself, then, as it sometimes happens, offered the most promising intermediate means of overtaking the chase. Accordingly, the boats now made for her, and were soon swayed up to their cranes—the two parts of the wrecked boat having been previously secured by her —and then hoisting everything to her side, and stacking her canvas high up, and sideways outstretching it with stun-sails, like the double-jointed wings of an albatross; the Pequod bore down in the leeward wake of Moby Dick. At the well known, methodic intervals, the whale's glittering spout was regularly announced from the manned mast-heads; and when he would be reported as just gone down, Ahab would take the time, and then pacing the deck, binnacle-watch in hand, so soon as the last second of the allotted hour expired, his voice was heard—"Whose is the doubloon now? D'ye see him?" and if the reply was, No, sir! straightaway he commanded them to lift him to his perch. In this way the day wore on; Ahab, now aloft and motionless; anon, unrestingly pacing the planks.

As he was thus walking, uttering no sound, except to hail the men aloft, or to bid them hoist a sail still higher, or to spread one to a still greater breadth—thus to and fro pacing, beneath his slouched hat, at every turn he passed his own wrecked boat, which had been dropped upon the quarter-deck, and lay there reversed; broken bow to shattered stern. At last he paused before it; and as in an already over-clouded sky fresh troops of clouds will sometimes sail across, so over the old man's face there now stole some such added gloom as this.

Stubb saw him pause; and perhaps intending, not vainly, though, to evince his own unabated fortitude, and thus keep up a valiant place in his captain's mind, he advanced, and eyeing the wreck exclaimed—"The thistle the ass refused; it pricked his mouth too keenly, Sir; ha! ha!"

"What soulless thing is this that laughs before a wreck? Man, man! did I not know thee brave as fearless fire (and as mechanical) I could swear thou wert a poltroon. Groan nor laugh should be heard before a wreck."

"Aye, Sir," said Starbuck drawing near, "'tis a solemn sight; an omen, and an ill one."

"Omen? omen?—the dictionary! If the gods think to speak outright to man, they will honorably speak outright; not shake their heads, and give an old wives' darkling hint.—Begone! Ye two are the opposite poles of one thing; Starbuck is Stubb reversed, and Stubb is Starbuck; and ye two are all mankind; and Ahab stands alone among the millions of the peopled earth, nor gods nor men his neighbors! Cold, cold—I shiver!—How now? Aloft there! D'ye see him? Sing out for every spout, though he spout ten times a second!"

The day was nearly done; only the hem of his golden robe was rustling. Soon, it was almost dark, but the lookout men still remained unset.

"Can't see the spout now, Sir;—too dark"—cried a voice from the air.

"How heading when last seen?"

"As before, Sir,—straight to leeward."

"Good! he will travel slower now 'tis night. Down royals and top-gallant stun-sails, Mr. Starbuck. We must not run over him before morning; he's making a passage now, and may heave-to a while. Helm there keep her full before the wind!—Aloft! come down!—Mr. Stubb, send a fresh hand to the foremast head, and see it manned till morning."—Then advancing towards the doubloon in the main-mast—"Men, this gold is mine, for I earned it; but I shall let it abide here till the White Whale is dead; and then, whosoever of ye first raises him, upon the day he shall be killed, this gold is that man's; and if on that day I shall again raise him, then, ten times its sum shall be divided among all of ye! Away now!—the deck is thine, Sir."

And so saying, he placed himself half way within the scuttle, and slouching his hat, stood there till dawn, except when at intervals rousing himself to see how the night wore on.

—Herman Melville,
Moby Dick, or the White Whale (1851)

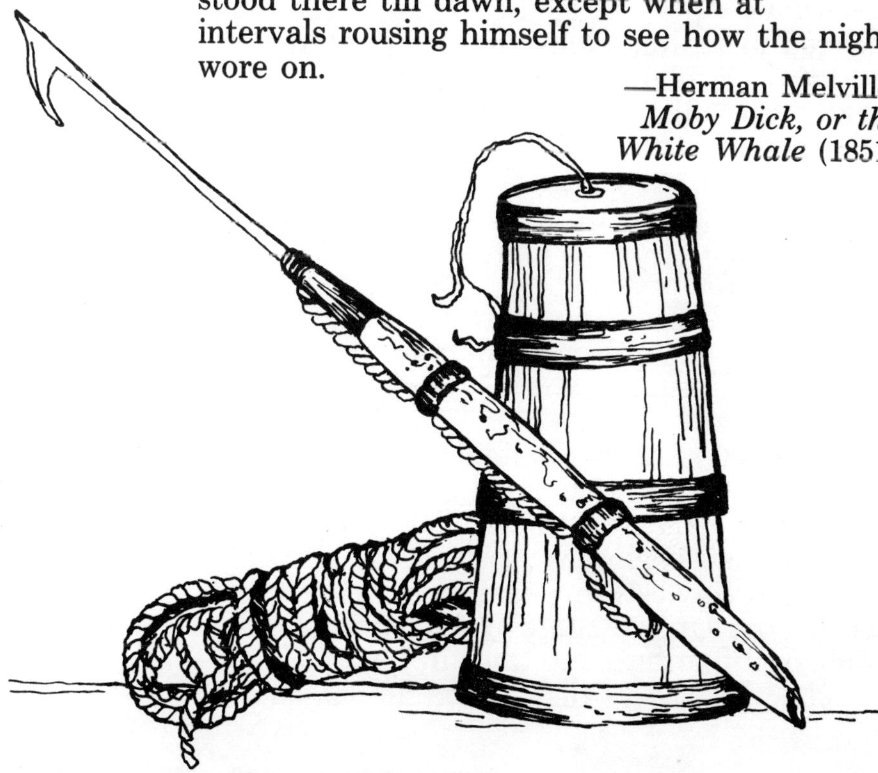

The Golden Age of Whaling

Charles M. Scammon

. . . While the explorations and the chase for both the Cachalot and the Right Whale were being vigorously prosecuted in the North and South Atlantic, and through the temperate and torrid zones, not only by American whalemen, but by vessels wearing the flags of the principal maritime nations of Europe, those remote and forbidding latitudes of the Indian Ocean and the South Pacific had received due attention. As far back as 1803, ships were cruising around Kurguelen Land for Right Whales, in the season, and sometimes a portion of their crews were engaged in sealing along the surf-beaten shores of Desolation and the Crozet islands, making up "mixed" but profitable voyages. Subsequently the coasts of New Zealand and New Holland (now Australia), became prolific whaling-grounds. Yet, with all the vast extent of both sea and ocean known to the whalemen for prosecuting their vocation, there were adventurous spirits among them who were ever in deep study and eager for a new field of pursuit, and plying their vessels to the far north in the Pacific, and unparalleled success awaited them. In the year 1835, the American ship *Ganges* took the first Right Whale on the Kodiak Ground. This was the beginning of the great whaling of the North-western Coast; and in 1839 the fleet of the United States engaged in whaling numbered five hundred and fifty-seven vessels, which were distributed among the Northern Atlantic ports in the proportions set forth in the subjoined table. In 1842 the number was six hundred and fifty-two. At this time the foreign whaling-fleet amounted to two hundred and thirty sail, and the combined fleet of the world, engaged in the enterprise, numbered eight hundred and eighty-two ships, barks, brigs, and schooners.

In 1846, the American force engaged was six hundred and seventy-eight ships and barks, thirty-five brigs, and twenty-two schooners, aggregating two hundred and thirty-three thousand one hundred and eighty-nine tons, valued at $21,075,000. At the same time, all the investments connected with the business are said to have been at least $70,000,000, and seventy thousand persons derived their chief support from the whaling interests. The first Bowhead Whales taken in the North Pacific, or in Behring Sea, were by the American ships *Hercules*, Captain Ricketson, and *Janus*, Captain Turner, which were on the coast of Kamschatka in 1843. In 1847, Bowheads were discovered in the Okhotsk Sea; and in 1848, Captain Royce, in the bark *Superior*, of Sag Harbor, was the first to pass through Behring Strait, and capture the *Baloena mysticetus* of the Arctic Ocean. From 1846 to 1851, inclusive, the

The Golden Age of Whaling
From "The American Whale-Fishery," in *Marine Mammals of the North-Western Coast of North America* (San Francisco, 1874; reprinted 1968 by Dover Publications, Inc.).

The American Whaling Fleet, 1839

Places where owned	Ships and Barks	Brigs and Schrs.	Amount of Tonnage	Places where owned	Ships and Barks	Brigs and Schrs.	Amount of Tonnage
New Bedford	169	8	56,118	Portsmouth	1	..	348
Fairhaven	43	1	13,274	Newport	9	2	3,152
Dartmouth	3	..	874	Bristol	5	1	1,782
Westport	5	4	1,443	Warren	18	3	6,075
Wareham	2	2	904	Providence	3	..	1,086
Rochester	5	10	2,615	New London	30	9	11,447
Nantucket	77	4	27,364	Stonington	7	5	2,912
Edgartown	8	..	2,659	Mystic	5	3	1,797
Holmes' Hole	3	1	1,180	Sag Harbor	31	..	10,605
Fall River	4	3	1,604	Greenport	4	1	1,414
Lynn	4	..	1,269	New Suffolk	1	..	274
Newburyport	3	..	1,099	Jamesport	1	..	236
Plymouth	3	..	910	Bridgeport	3	..	913
Salem	14	..	4,265	New York	3	..	710
Boston	..	1	125	Hudson	8	..	2,902
Dorchester	2	..	581	Poughkeepsie	6	..	2.043
Falmouth	8	..	2,490	Cold Springs	2	..	629
Provincetown	..	1	172	Wilmington	5	..	1,578
Portland	1	..	388	Newark	1	..	366
Wiscasset	1	..	380	....	..	..	....

whale-fishery of the United States may be regarded as having been in its most flourishing condition, the average number of vessels annually employed for these years being six hundred and thirty-eight, with an aggregate tonnage of two hundred and two thousand two hundred and seventy-two tons. This immense fleet was scattered around the world. . . .

—Charles M. Scammon (1874)

The Whale Fishery and Its Products (1915)

The chief products of the fishery are sperm and whale oil, whalebone, and ambergris. Spermaceti, meaning a foot of "sperm oil," was the most valuable and was found only in the sperm whale. This oil was formerly used chiefly in the manufacture of sperm candles, and at one time there were eight factories for the manufacture of these candles in New England, Nantucket alone turning out three hundred and eighty tons annually before the war. In the olden times this oil was considered a sure cure for almost any kind of disease and was worth its weight in silver. Shakespeare makes reference to it in these words—"The sovereign'st thing on earth was 'parmaceti for an inward bruise." At present it is used chiefly in making refined oils for lubricating.

Whale oil was procured from all the other varieties of whales, and was formerly used as an illuminant in the old "whale oil" lamps; it is used now to a certain extent in the tanning of leather and in the manufacture of soaps, but chiefly in making heavy lubricating oils.

Whalebone has been the most important product of the whale fishery for a number of years, and in fact whaling would undoubtedly have died out altogether had it not been for the discovery of its use in making women's stays. Many a whaleman has lost his life in the endeavor to improve the female figure. It is a curious fact that fifty years or more ago this product was always thrown away as worthless. The value has gone down in the past few years on account of the invention of steel stays, which take the place of whalebone.

The high and low prices of these three commodities are of interest. Sperm oil was $2.25 per gallon in 1866, and is 46 cents now [1915]. Whale oil was $1.45 per gallon in 1865, and is 26 cents now. Whalebone was $5.80 per pound in 1904, 8 cents in 1809, and is $1.75 now.

Ambergris, the rarest and most valuable of all the products, is a secretion from the intestines of the sperm whale and results from a disease. It is a very rare article and is worth almost its weight in gold, selling usually at $300 a pound. Its chief use is in the preparation of fine perfumeries. It is believed that the largest amount taken by one ship was brought back by the "Watchman" of Nantucket, which vessel found eight hundred pounds in 1858. Small amounts were sold every year in New Bedford even up to the year 1913. The Turks used it in cooking and also carried it to Mecca for the same purpose that frankincense is carried to St. Peter's in Rome. Some wine merchants used to drop a little into their wine as a spice, and it was said that the Moors used it in green tea as a flavoring to present to their guests.

The whale is used for food chiefly by the Japanese and Esquimaux, and a famous

38 *The* **Charles W. Morgan** *last of the wooden whalers, was built in New Bedford in 1841 and whaled for more than 80 years. (Photo courtesy of Mystic Seaport, Inc.)*

doctor belonging to the latter tribe some years ago recommended the blubber for infants. In fact, the whale would perhaps be considered a good dish were there not so much of him. Whalemeat is said by some to resemble boarding-house steak. In France, during the Middle Ages, the tongue was considered a great delicacy, and by some epicures the brains, mixed with flour, were much sought after.

The largest income received by the whalers of America in any one year was in 1854, when they netted $10,802,594.20, although the record size of the fleet was attained eight years before. The five years from 1853 to 1857 inclusive yielded a return of $51,063,659.59, the catch of each year selling for fifty per cent of the total value of the whaling fleet. The total value of the cargoes from 1804 to 1876 was $331,947,480.51.

Captain W.T. Walker, of New Bedford, is called the counting-house hero of the American Whale Fishery. He purchased in 1848 an old whaleship called the *Envoy* that was about to be broken up, and when ready for sea this ship stood the owner $8,000. He could get no insurance; nevertheless he "took a chance," and after a three years' voyage he returned and had netted for himself the extraordinary sum of $138,450, or 1,630 per cent. The largest profit, however, was made by the *Pioneer* of New London, in 1865, the value of her cargo being $151,060. For a short voyage Frederick Fish, who has been mentioned before, holds the record for his ship the *Montreal*, which brought back a cargo worth over $36,000 after a voyage occupying only two months and fifteen days.

There were many unprofitable voyages, and many were the ships that came home with barrels filled with salt water instead of oil for ballast. Some vessels, as whalemen say, didn't have enough oil to grease their irons.

—*Whale Fishery of New England—An Account, With Illustrations and Some Interesting Anecdotes, of the Rise and Fall of an Industry Which Has Made New England Famous Throughout the World* (1915)

Whale Watchers: King Alfred the Great of England (c. A.D.890)

Octher told his lord king Alfred that he dwelt furthest North of any other Norman. . . . He said that upon a certeine time he fell into a fantasie and desire to proove and know how farre that land stretched Northward, and whether there were any habitation of men North beyond the desert. Whereupon he tooke his voyage directly North along the coast, having upon his steerboord alwayes the desert land; and upon the leereboord the maine Ocean: and continued his course for the space of 3 days. In which space he was come as far towards the North, as commonly the whale hunters use to travell. Whence he proceeded in his course still towards the North so farre as he was able to saile in other 3 dayes. . . . and thence he sailed plaine East along the coast still so far as he was able in the space of 4 dayes. . . he sailed thence along the coast continually full South, so farre as he could travaile in 5 dayes; and at the fifth dayes end he discovered a mightie river which opened very farre into the land. . . .The principall purpose of his traveile this way was to encrease the knowledge and discoverie of these coasts and contreyes, for the more commoditie of fishing of horse-whales, which have in their teeth bones of great price and excellencie: whereof he brought some at his returne unto the king. Their skinnes are also very good to make cables for shippes and so used. This kinde of whale is much lesse in quantitie then other kindes, having not in length above seven elles. And as for the common kind of whales, the place of most and best hunting of them is in his owne countrey: whereof some be 48 elles of length, and some 50 of which sort he affirmed that he himselfe was one of the sixe, which in the space of 3 dayes killed three-score. He was a man of exceeding wealth in such riches, wherein the wealth of that countrey doth consist. . . . Their principall wealth consisteth in the tribute which the Fynnes pay them, which is all in skinnes of wilde beasts, feathers of birds, whale bones, and cables and tacklings for shippes made of Whales or Seales skinnes.

From "The Voyage of Octher," by King Alfred the Great as recounted by King Alfred in his *Description of Europe,* and translated by Hakluyt in his *Principal Navigations.*

Scrimshaw: The Whaler's Art

Scrimshaw is the art of engraving on whale bone, whale's teeth, elephant tusks or bit of ivory. It is essentially a primitive folk art originating with the sailors on American whaling ships of the 19th century. Early examples are avidly sought by knowledgeable collectors. President John F. Kennedy was a connoiseur of scrimshaw — one of his cherished pieces, a sperm whale's tooth engraved with the Presidential seal, was buried with him at the direction of his wife.

Some of these patiently and laboriously wrought pieces pinpointing the daily life of a lonely sailor are of great grace and beauty, but in most cases their charm lies in a sincere and personal statement of a way of life now gone.

In these days of unlimited "limited" editions of all sorts, we should, perhaps, point out that these precious artifacts, whether lovingly created on a rolling or pitching ship in whale waters over a hundred years ago or by a devoted modern artist working in the comfort of his studio-den lined with pictures, books, and seafaring relics, is a one-of-a-kind piece of original art.

—Martin David Dubin

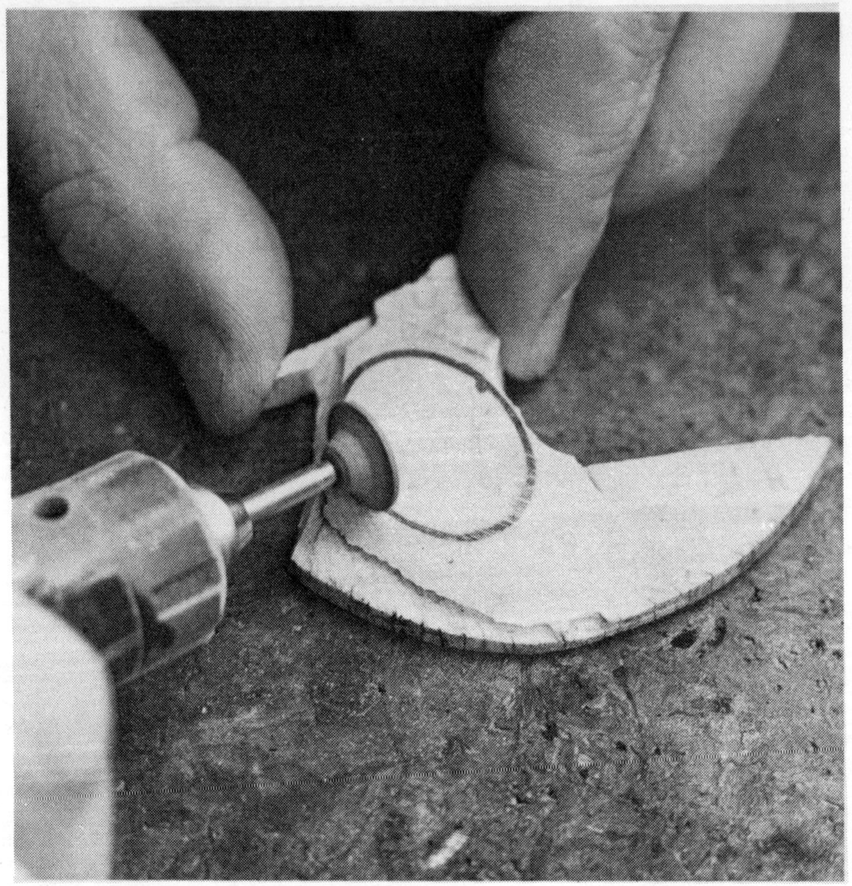

39
A skilled scrimshander (someone who works in scrimshaw) practicing his trade. (Photo courtesy of Martin David Dubin)

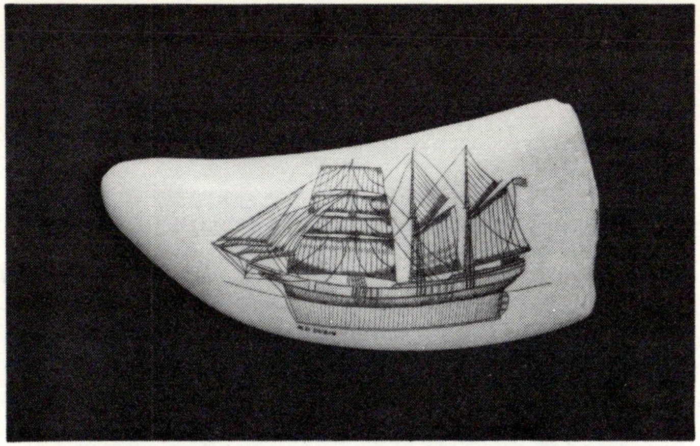

40
Although the large pieces, left, are engraved whale teeth, many scrimshanders now use elephant ivory for art. (Photo courtesy of Bob Porth, Hedrich-Blessing)

Plate XXVI.

C. M. Scammon del.

A NORTHERN WHALING SCENE.

Britton & Rey, lithos.

41
A Northern whaling scene. (Drawings from Scammon's **Marine Mammals of the North-Western Coast)**

The Greenland Fishery

"The Greenland Fishery" was popular in the English, Scottish, and American whaling fleets and in the merchant marine as well; it was even circulated as a broadside ballad in the early 1800s. This seems to be one of the earliest versions of this forecastle song.

It was in the year eighteen hundred and one,
On March the twentieth day,
Our sheets were spread, and our anchors weighed,
And for Greenland we bore away, brave boys,
And for Greenland we bore away.

Speed it was our Captain's name,
And our ship the *Lion* bold,
And we sailed away to that cold countrie,
For to face the storms and the cold, brave boys,
For to face the storms and the cold.

And when we came to that cold countrie,
Where the frost and the snow doth lie;
The frost and the snow and the wind doth blow,
And the daylight never away, brave boys,
And the daylight never away.

We had harpooners and spectioneers
And line-coilers also,
We had men to row and men to tow,
To the whale where'er that she goes, brave boys,
To the whale where'er that she goes.

The captain has gone to the topmast high,
With a spy-glass in his hand,
A whale, a whale, a whale, cries he,
And she blows at every span, brave boys,
And she blows at every span.

Our mate stood on the quarter-deck,
And a quick little man was he,—
"Overhaul, overhaul, let the boat-tackle fall,"
And we launched them into the sea, brave boys,
And we launched them into the sea.

The whale was struck, and away she went,
With a flourish of her tail.—
Alas, alas! and we lost four men,
But we did not catch the whale, brave boys,
But we did not catch the whale.

And when the news to the captain was brought,
He down his colours drew,
Crying, alas, alas, for my four pretty men,
The darlings of our crew, brave boys,
The darlings of our crew.

But what needs I regret and cry
For the losing of my hands?
For fortune will be fortune still,
Let a man do all that he can, brave boys,
Let a man do all that he can.

But as long as the water runs below the bridge
And the sea doth ebb and flow,
We will go no more to Greenland, brave boys,
We will go no more to Greenland, brave boys,
We will go no more, oh no!

42
(Photo courtesy of NOAA)

6

THE LEGAL WHALE

. . . Survival chances for the human race will be greatly enhanced when man concedes to the Earth and all its life forms the right to exist that he wants for himself. The only homage he can now pay to the blue whale is to learn the lessons of dependence on and kinship with all life. If he does not learn them, the great blue whale will have died in vain—having taught nothing to his only mortal enemy.

— *George B. Small,*
The Blue Whale

Watching Out for the Whales:

I. A Look at the IWC Ken Hampton

The International Whaling Commission (IWC) is the self-regulating association of whaling nations. Meeting annually, the Commission sets quotas on the numbers of each species of whale that can be taken and bans the killing of other species. Some member nations, such as the United States, no longer engage in commercial whaling and have a moderating influence on the whaling nations.

There are, however, several drawbacks to the IWC:

They have no enforcement powers over member nations. The IWC is powerless to enforce compliance with the whale quotas. In recent years observers have been allowed on board whaling ships to report on the catch, but this is far from an effective system. At present, growing public sentiment is the only effective means for pressuring whaling nations to comply with the quota system.

Not all whaling nations are members. South Korea, Spain, Portugal, Chile, Peru and Somalia are outside even the limited control of the IWC, although responsible for approximately 10 percent of the annual kill.

IWC is the only source of data on whale population estimates, and it is these estimates that are the basis for setting the quotas for each species of whale. Whalers are not trained to collect scientific data and it is in their best interest to keep population estimates high. Quotas over the last several years have been lowered, in part, because whale stocks continue to diminish and because of improved management procedures initiated by the United States.

Comparison of Catch Limits Set by IWC

1973	37,500	1976	27,939
1974	37,300	1977	(to come)
1975	32,578		

Under IWC regulations, some species are completely protected, including the blue, gray, humpback, right and bowhead whales. Having reached the level of "commercial extinction" due to extensive whaling, their numbers are so diminished that it is no longer profitable to hunt them. The California gray whale, however, has increased in number, while the blue whale, for example, shows no sign of population growth despite more than ten years of protection.

Changes in IWC policies and quotas since 1974 indicate that they do respond to the growing pressure of the conservation movement. . . . As more and more people become aware of the plight of the whales and become involved in the efforts to save them, the better the chances that whales will survive in future generations.

Watching Out for the Whales: A Look at the IWC
From *The Whale Report,* Center for Environmental Education.

Watching Out For The Whales: II. Good News for the Whale (1976)

The author is the assistant conservation director of the National Wildlife Federation and was a member of the U.S. delegation to the 1976 IWC meeting.

Sperm whale quotas in the Southern Hemisphere were cut by more than 50 percent to 4,791 (from last year's 10,740), while quotas established for whales in all oceans were reduced by nearly 6,000 from last year's total catch limit of 32,000 animals. Also, all whale stocks were brought under international regulation for the first time and the regionalization of quotas was further refined. Those were some of the most significant actions taken by the 16-member International Whaling Commission at its 28th annual meeting (June 1976) in London, England.

The reduction in catch limits of almost 6,000 whales is especially important because it involves reduced quotas on the largest whales, such as the sperm, fin, sei, and Bryde's. The new worldwide quotas are the lowest in the history of the Commission. Only in the case of the minke whale, the smallest of the so-called great whales and a species relatively unexploited thus far, has the quota been increased.

Currently, the IWC recognizes five species of whales as being endangered and has afforded total protection to them: blue, gray, humpback, right, and bowhead. The U.S. considers three additional species as endangered: sperm, fin, and sei.

Member Nations of IWC

Commercial Whaling Nations:

Australia	Japan
Brazil	Norway
Canada	South Africa
Denmark	Soviet Union
Iceland	

Non-Whaling Nations:

Argentina	New Zealand
France	United Kingdom
Mexico	United States

. . . According to a Commission ruling, an IWC member has up to 90 days to officially advise the Secretary if it decides not to conform with the quotas established in plenary session (by three-quarters majority vote) at the Commission's annual meetings. This means that both Japan and Russia, which together account for more than 80 percent of the whales harvested annually, will have until the end of September [1976] to make a decision. Official silence on the subject is considered acceptance of the IWC-established catch limits.

The USSR will probably feel the greatest impact from the reduced quota on sperm whales because of its apparent continued reliance on sperm oil for use in missiles and high-precision instruments. Most nations, including the U.S., have found satisfactory synthetic substitutes for sperm oil and for

---

Watching Out for the Whales: Good News for the Whale (1976)
From *Conservation News,* August 15, 1976.

whale parts, formerly used in the manufacture of such items as cosmetics and car wax.

Japan insists that it must continue commercial whaling because whale meat is an important national source of animal protein. The Japan Whaling Information Center claims that whales provide approximately nine percent of the meat consumed in Japan. Many outsiders point out, however, that whale meat actually constitutes less than two percent of Japan's total protein intake, most of which comes from fish and shellfish, as well as soybeans and animal meat. Such critics contend that Japan, for economic reasons, is simply attempting to perpetuate the anachronistic whaling industry. They acknowledge, though, that both Japan and the USSR have made significant reductions in the number and size of their ocean whaling fleets, especially within the past year. [The Russians reduced their fleets from three to two; while Japan consolidated the whaling divisions of six companies, halving the work force and reducing their whaling fleets from a total of six to one in the North Pacific and two in the Antarctic.]

The International Whaling Commission was once dismissed by many conservationists as an ineffectual international body whose only apparent purpose was to lend a cloak of legality to the whaling industry's over-harvest of great whales. But the IWC . . . appears to be slowly changing its character and its approach to whale exploitation.

In June 1975 the Commission took affirmative action in adopting a new "management procedure" which, in effect, established a selective moratorium on the commercial taking of great whales. The procedure calls for the total protection of any stock of a species that falls 10 percent or more below the so-called Maximum Sustainable Yield (MSY) level, as estimated by the foremost whale experts from IWC member nations who comprise the Commission's Scientific Committee . . . .

The U.S. and several other IWC members would prefer to have the IWC adopt a ten-year total moratorium on commercial whaling as recommended in 1972 at the UN Conference on the Human Environment at Stockholm. However, the U.S. Commissioner to the IWC, NOAA Administrator Dr. Robert M. White, and a few other concerned commissioners have been unable, thus far, to garner anything approaching the three-quarters majority vote necessary.

In approving its Scientific Committee's recommendations, calling for sharply reduced sperm whale quotas in the Antarctic, the IWC served notice on Japan and Russia that it intends to support the management procedure developed by its scientists . . . and strengthen the Commission's program of whale conservation despite considerable pressure from Japan, the USSR, and a few other whaling nations to permit larger quotas. Thus, it seems that the IWC, despite all its shortcomings, is the most viable international mechanism for safeguarding the welfare of great whales.

—*Ken Hampton*

43
Pilot whale. (Photo courtesy of Randall R. Reeves)

The Rights of Cetaceans Under Human Laws

Dr. John Lilly, M.D.

The rights of man have slowly developed over the last few centuries. The rights that each of us enjoys today in the United States have been carefully developed in our laws. Our past history teaches us that our rights evolved through certain stages, from the unconscious acceptance of a lack of rights, to the conscious awareness of the need for an adequate expression of an intolerable situation or state, to the demand for the relief articulated in law, to the law and its adequate administration. As each group of humans, through its own experience, learned to feel its lack of sharing in the benefits of laws and their administration, each developed adequate spokesmen or spokeswomen for its cause. These individuals were either inside or outside the group needing relief.

Professor Christopher D. Stone, in *Should Trees Have Standing* , has given an excellent review of the origins of these laws for human rights and shows the extension of the legal concepts to include fictitious individuals. In these laws, the concept of an individual with legal rights is extended to corporations, cities, states and nations. Professor Stone espouses extending the concept of an individual with legal rights to the environment including animals. In such cases, individual humans and groups must become the legal guardians or protagonists for the environment and the animals. Since trees, for example, cannot speak in human language, persons must express the need of trees for rights under human law.

What is the need for relief of an intolerable situation for the cetaceans? Should cetaceans be given individual rights under human laws? Are there, as yet, enough informed, consciously aware humans to express the needs of cetaceans, articulate them in legal terms, formulate the laws and assure their passage in the legislature and courts? Is the nature of cetaceans such that they warrant special laws?

The plight of the large whales is extensively documented by the International Whaling Commission and several interested biologists, such as Roger Payne, Victor Scheffer and Scott McVay. The extinction of several species, including the largest animal the earth has ever housed, the blue whale, has apparently already occurred. The whaling industry kills a whale every fifteen minutes of every day. The survival of the entire species and genera is threatened. "Specicide" is a word to add to our laws.

Dolphins, including the largest one *Orcinus Orca* , the so-called "killer-whale" (a very curious name for this highly sensitive species. In the last century the term was "whale killer." Their record is clean: no orca has

---

The Rights of Cetaceans Under Human Laws
Reprinted from *Oceans* magazine, a publication of the Oceanic Society, © 1976; by permission of the publishers and Dr. Lilly.

44 Blue whale at Marine Land, Venice, California. (Photo courtesy of U.S. Navy)

been recorded as ever having attacked a human who had not provoked attack), also have an intolerable situation needing relief. In Japan the dolphins are killed for human use and food. In numerous countries around the world they are captured for human entertainment and kept in confined quarters, restricted beyond their natural limits, and forced to do repetitious circus performances of a peculiarly banal nature. They are cut off from communication with their relatives and friends in the sea. Many die prematurely during capture and from the confinement.

Large numbers of people have recently become aware of the whales' situation and are asking for cetacean relief under our laws. Recently the Greenpeace Foundation obtained a ruling from the Canadian Government prohibiting further capture of *Orcinus orca* in the waters of Canada. Their collected data on the deaths of these creatures during capture, transport and confinement apparently were the deciding factors.

Are cetaceans something more than animals whose survival is in jeopardy through human actions unrestrained by law? Is their nature such that we should seek more vigorous legal action to preserve their way of life, their territory? Are they in any sense special groups needing special attention from us?

Let us state the question in scientific and humanistic terms. What is the probability that cetaceans are sentient, intelligent, creatures?

Consider their brains and our brains. How are we, in the structure of our brains, different from our nearest cousins, the apes (chimpanzee, gorilla and orangutang)? The result of careful neuroanatomical and extensive neurophysiological studies reveal that our brain's only decisive difference from the apes' is in the size of our cerebral cortical "silent" areas in the frontal, parietal and temporal lobes. Silent areas have no direct-input connections and thus are devoted to central processing (thinking, imagination, long-term goals, ethics, etc.). Without such areas, a person is no longer a human as each of us conceives of being human. Persons in whom these areas are lost become here-and-now beings, with a good deal of their essential humanness gone. They lose their moral and ethical judgments and their motivations for future planning and action. These cortical areas are the ones we use for understanding justice, compassion and the need for social interdependence. If we could control the growth of these cortical areas in our species, we could possibly evolve further beyond our present horizons. Apparently, we are limited by this aspect of our brain's structure. This may make us pause when we consider the brains of *Cetacea.*

Recent, excellent, controlled neurological studies of cetacean brains by Morgane, Yakolev, Jacobs and several Russian scientists show that the silent areas are larger in the cetaceans whose brains are bigger than ours. In the bottlenose dolphin, with a brain forty percent larger than ours, in *Orcinus orca,* with a brain three to four times the mass of ours and in the sperm whale, whose brain is the largest on this planet (six times the mass of a human brain), all of the additional mass is in the silent areas.

The size of these cetacean brains has been known for many years. Only recently has microscopic analysis shown that their cellular densities and connections are quite as large and complex as ours.

The belief that man is the preeminent thinker, doer, feeler on this planet interferes with these investigations. As fast as evidence of the superior nature of cetacean brains accumulated, rationalizations were devised (even by neurologists) to place these brains in a category below man's. Two main arguments have been used against acceptance of the probability of brains superior to man's.

The first rationalization is that of body size, or weight, versus brain size, or weight. This argument holds that a large body needs a large brain to control its behavior. In the light of the findings on cetacean brains, this argument is not relevant. The body is controlled by portions of the brain outside the silent areas. Moreover, the size of these outside areas is not proportional to the size of the individual animal. In our own case the sensory and motor areas of the cortex are no larger than those of the chimpanzee or the gorilla. In the case of sea creatures, the whale shark (not a mammal), with a body weight of forty tons, has a brain mass of 100 grams, less than that of a macaque monkey. A forty-ton sperm whale has a brain mass of 9,000 grams, six times ours.

The second argument for the preeminence of man over larger brained cetaceans points to the human accomplishments of using hands, planning structures and building them. Cetaceans have no hands and therefore had no need to develop intelligence.

This argument is derived from man's own narcissistic need to see the use of intelligence and sentience only in areas most men have entered. May there not be domains only a few men have penetrated in which whales may be superior? May not their philosophies and traditions be more complex, more full of insight than ours? Cut off from the need for building, for external forms of transport, for food preservation and preparation, they probably have very advanced ethics and laws, developed over millennia and passed on, through sonic communication, to their young. Their memories are probably much greater than ours. Like those among us with no written language (the Masai, for example), they probably have long histories which they recount to their young, who must memorize them in detail.

Who are to specify the criteria for intelligence among large-brained mammals of the sea? We have no direct experience of living in the sea. We venture timidly into water protected by articial means. Let us try to find out from the *Cetacea* their criteria for intelligent use of large brains, rather than attempting to impose our criteria upon them.

What behavioral evidence do we have of ethics among *Cetacea* ? The main evidence is derived from our encounters with them, and some from our observátions of their interactions with one another. In the books *Man and Dolphin, The Mind of the Dolphin (a non-human intelligence),* and *Lilly on Dolphins: the Humans of the Sea,* are many examples of intelligent and compassionate actions between us and them, and between individual dolphins. I shall not repeat those accounts here, for I could not do them justice in this brief article.

However, the main points made in these books are as follows:

(1) The brains of cetaceans are superior to ours in unique and effective areas.

(2) The behavior of *Cetacea* in captivity is regulated by a cetacean education and a cetacean ethic in which man is carefully treated as a special case. Man is not to be injured by an individual cetacean even under extreme degrees of provocation. Even "cruel and unusual punishment" shall not be responded to with violent destructive action on the bodies of humans. Throughout their capture and during their performances, men

45 Pilot whale in Navy research "deep operations recovery" program. (Photo courtesy of the U.S. Navy)

and women are in direct contact with them, and none have been injured by these huge creatures.

(3) Dolphins treat their own individuals with total knowledge of the necessities for survival of the group. Individuals will commit suicide if their incapacity becomes great enough to endanger the group. Any sick or grieving dolphin is cared for by the group; but if the care interferes with group survival, the individual voluntarily stops breathing and thus commits suicide.

(4) Dolphin sonic-ultrasonic communication is more complet than ours, ten times as fast, at ten times the frequency.

It is highly probable that their minds operate with acoustic analogs, even as our language is primarily based on visual analogs. They "see" their environment and one another with sound. Their sonar "sees" into both our bodies and their own. They can see emotional states in one another through stomach and lung movements. They become acquainted with, and subsequently recognize, one another and humans in the water with them, through the transmission of sonic pictures.

Each dolphin has three independently controlled sonic-ultrasonic emitters (two nasal and one laryngeal). Two dolphins exchanging information may use all three separate channels at once.

(5) Dolphins are interested enough in communicating with us for them to exert surprising efforts to reprogram their outputs to solve communication tasks imposed by us.

(6) It is probable that sufficiently fast computers could enable man to communicate with dolphins and whales.

Such considerations inevitably lead us to the conclusion that *Cetacea* are a special case. We must give them rights as individuals under our laws. The following guidelines for new laws are suggested:

(1) Cetaceans are no longer to be considered as property, nor as an industrial resource, nor as stocks of animals.

(2) Cetaceans are to have complete freedom of the waters of the earth.

(3) Individual dolphins and whales are to be given the legal rights of human individuals. Human individuals and groups are to be given the right to sue in behalf of, or otherwise represent in court, cetacean individuals placed in jeopardy by other humans.

(4) Research into communication with cetaceans is no longer simply a scientific pursuit: such research is now a necessity for people to exchange information at a high level of complexity with cetaceans. We must learn their needs, their ethics, their philosophy, to find out who we are on this planet, in this galaxy. The extraterrestrials are here—in the sea. We can, with dedicated efforts, communicate with them. If and when we break the communication barrier, then we and the cetaceans can work out our differences and our correspondences.

We may already have accumulated too negative a reputation among cetaceans. Let us at least stop killing and enslaving them for our entertainment and our warfare. This we can do now. The communication problem will take longer and requires much research.

In the Human/Dolphin Foundation we are collecting a bibliography on dolphins and whales in the areas of brain structure, communication and behavior. The Foundation is also proposing research to start breaking the communication barrier between man and the dolphin. The research is to be conducted at sea with no confinement of dolphins.

—*Dr. John Lilly, M.D.*

Whale Watchers: The California State Legislature

In 1953 California adopted the grizzly bear as its state animal — 42 years after it had become extinct in the state.

This summer (1975) the state legislature adopted the California gray whale as the official state marine mammal. And it is hoped that the gray whale's future will turn out differently than the grizzly that adorns the California state flag.

States which have long designated official state birds, flowers, animals, trees, and even insects and reptiles are beginning to add whales to their pantheon of official symbols.

Connecticut recently followed California's lead by designating the sperm whale as its official state animal. . . .

—Todd Crowell

By Todd Crowell in the *Christian Science Monitor,* September 30, 1975.

EPILOGUE

PORT OF MANY SHIPS

John Masefield

"Down in the sea, very far down, under five miles of water, somewhere in the Gulf of Mexico, there is a sea cave, all roofed with coral. There is a brightness in the cave, although it is so far below the sea. And in the light there the great sea-snake is coiled in immense blue coils, with a crown of gold upon his horned head. He sits there very patiently from year to year, making the water tremulous with the threshing of his gills. And about him at all times swim the goggle-eyed dumb creatures of the sea. He is the king of all the fishes, and he waits there until the judgement day, when the waters shall pass away for ever and the dim kingdom disappear. At times the coils of his body wreathe themselves, and then the waters above him rage. One folding of his coil will cover a sea with shipwreck; and so it must be until the sea and the ships come to an end together in that serpent's death-throe.

"Now when that happens, when the snake is dying, there will come a lull and a hush, like when the boatswain pipes. And in that time of quiet you will hear a great beating of ships' bells, for in every ship sunken in the sea the life will go leaping to the white bones of the drowned. And every drowned sailor, with the weeds upon him, will spring alive again; and he will start singing and beating on the bells, as he did in life when starting out upon a cruise. And so great and sweet will be the music that they make that you will think little of harps from that time on, my son.

"Now the coils of the snake will stiffen out, like a rope stretched taut for hauling. His long knobbed horns will droop. The golden crown will roll from his old, tired head. And he will lie there as dead as herring, while the sea will fall calm, like it was before the land appeared, with never a breaker in her. Then the great white whale, old Moby Dick, the king of all the whales, will rise up from his quiet in the sea, and go bellowing to his mates. And all the whales in the world—the sperm-whales, the razor-back, the black-fish, the rorque, the right, the forty-barrel Jonah, the narwhal, the hump-back, the grampus and the thrasher—will come to him, 'fin-out,' blowing their spray to the heavens. Then Moby Dick will call the roll of them, and from all the parts of the sea, from the north, from the south, from Callao to Rio, not one whale will be missing. Then Moby Dick will trumpet, like a man blowing a horn, and all that company of whales will 'sound' (that is,

---

Port of Many Ships
From *A Mainsail Haul* by John Masefield, first published 1905; reissued in the Mariners Library (London: Rupert Hart-Davis, 1954).

46
Bottoms up! (Photo courtesy of Randall R. Reeves)

dive), for it is they that have the job of raising the wrecks from down below.

"Then when they come up the sun will just be setting in the sea, far away to the west, like a ball of red fire. And just as the curve of it goes below the sea, it will stop sinking and lie there like a door. And the stars and the earth and the wind will stop. And there will be nothing but the sea, and this red arch of the sun, and the whales with the wrecks, and a stream of light upon the water. Each whale will have raised a wreck from among the coral, and the sea will be thick with them—row-ships and sail-ships, and great big seventy-fours, and big White Star boats, and battleships, all of them green with the ooze, but all of them manned by singing sailors. And ahead of them will go Moby Dick, towing the ship our Lord was in, with all the sweet apostles aboard of her. And Moby Dick will give a great bellow, like a foghorn blowing, and stretch 'fin-out' for the sun away in the west. And all the whales will bellow out an answer. And all the drowned sailors will sing their chanties, and beat the bells into a music. And the whole fleet of them will start towing at full speed towards the sun, at the edge of the sky and water. I tell you they will make white water, those ships and fishes.

"When they have got to where the sun is, the red ball will swing open like a door, and Moby Dick, and all the whales, and all the ships will rush through it into an anchorage in Kingdom Come. It will be a great calm piece of water, with land close aboard, where all the ships of the world will lie at anchor, tier upon tier, with the hands gathered forward, singing. They'll have no watches to stand, no ropes to coil, no mates to knock their heads in. Nothing will be to do except singing and beating on the bell. And all the poor sailors who went in patched rags, my son, they'll be all fine in white and gold. And ashore, among the palm-trees, there'll be fine inns for the seamen, where you and I, maybe, will meet again, and I spin yarns, maybe, with no cause to stop until the bell goes."

—John Masefield (1905)

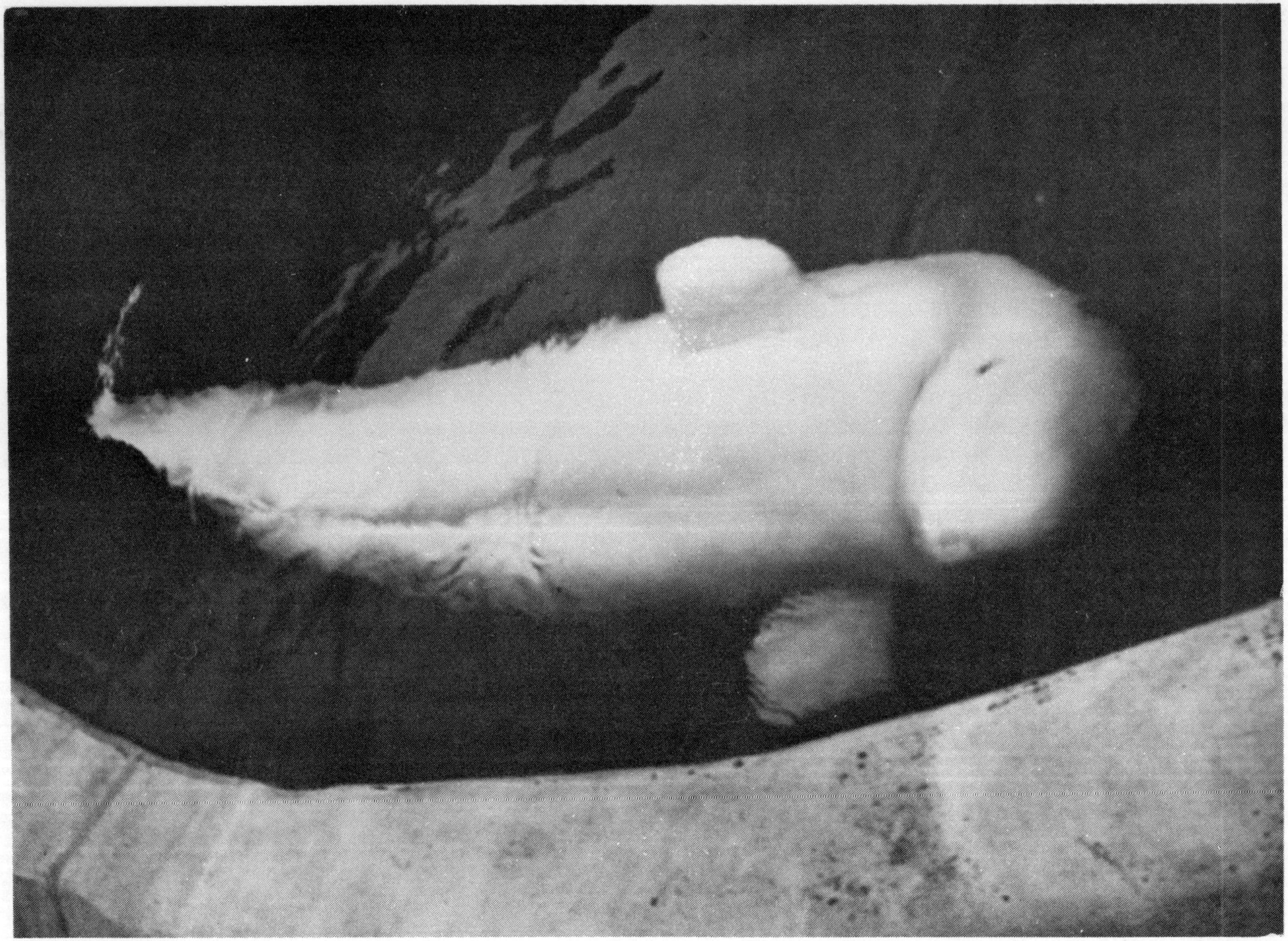

47
Beluga whale at New York Aquarium. (Photo courtesy of Randall R. Reeves)

Saving The Whales

A list of some projects and organizations concerned with protecting the great whales. (Courtesy of the John G. Shedd Aquarium, Chicago.)

American Cetacean Society
4725 Lincoln Blvd.
Marina Del Rey, Calif. 90291

Cousteau Society
P.O. Box 1000
Bridgeport, Conn. 06601

Committee on Whales
Environmental Defense Fund
162 Old Town Road
East Setauket, N.Y. 11733

General Whale
9616 MacArthur Blvd.
Oakland, Calif. 94605

Greenpeace Foundation
860 2nd St.
San Francisco, Calif. 94107

Whale Fund
New York Zoological Society
Bronx Zoo
185th St. & Southern Blvd.
Bronx, N.Y. 10460

Project Jonah
P.O. Box 476
Bolinas, Calif. 94924

Rare Animal Relief Effort (RARE)
c/o National Audubon Society
950 Third Ave.
New York, N.Y. 10022

Sierra Club Whale Project
530 Bush St.
San Francisco, Calif. 94108

U.S. Marine Mammal Commission
1625 I Street, NW (Room 1625)
Washington, D.C. 10006

Whale Protection Fund
Center for Environmental Education
2100 M Street, NW
Washington, D.C. 20037

INDEX